"Highly recomm   s
a perennial curre   e
wondering how cc _____ naturalism connect
with older materialist philosophies and politics."

William Lewis, Skidmore College, USA

"Brown and Ladyman offer a clear exposition of philosophical materialism much needed in these muddle-headed times. Particularly refreshing is their stress on the essential incompleteness of the explanations it provides, which distinguish it as a scientific worldview from its more strictly 'philosophical' rivals."

Thomas Uebel, University of Manchester, UK

Materialism

The doctrine of materialism is one of the most controversial in the history of ideas. For much of its history it has been aligned with toleration and enlightened thinking, but it has also aroused strong, often violent, passions amongst both its opponents and proponents. This book explores the development of materialism in an engaging and thought-provoking way and defends the form it takes in the twenty-first century.

Opening with an account of the ideas of some of the most important thinkers in the materialist tradition, including Epicurus, Lucretius, Hobbes, Hume, Darwin and Marx, the authors discuss materialism's origins, as an early form of naturalistic explanation and as an intellectual outlook about life and the world in general. They explain how materialism's beginnings as an imaginative vision of the true nature of things faced a major challenge from the physics it did so much to facilitate, which now portrays the microscopic world in a way incompatible with traditional materialism. Brown and Ladyman explain how out of this challenge materialism developed into the new doctrine of physicalism.

Drawing on a wide range of colourful examples, the authors argue that although materialism does not have all the answers, its humanism and commitment to naturalistic explanation and the scientific method is our best philosophical hope in the ideological maelstrom of the modern world.

Robin Gordon Brown is a Research Associate in the Department of Philosophy at the University of Bristol, UK.

James Ladyman is Professor of Philosophy at the University of Bristol, UK. He is the author of *Understanding Philosophy of Science* (2002) and editor (with Alexander Bird) of *Arguing About Science* (2012), both published by Routledge.

ROBIN GORDON BROWN
AND JAMES LADYMAN

Materialism
A Historical and
Philosophical Inquiry

Routledge
Taylor & Francis Group

LONDON AND NEW YORK

First published 2019
by Routledge
2 Park Square, Milton Park, Abingdon, Oxon OX14 4RN

and by Routledge
52 Vanderbilt Avenue, New York, NY 10017

*Routledge is an imprint of the Taylor & Francis Group,
an informa business*

British Library Cataloguing-in-Publication Data
A catalogue record for this book is available from the British Library

Library of Congress Cataloging-in-Publication Data
Names: Brown, Robin (Robin Gordon), author. | Ladyman, James, 1969- author.
Title: Materialism : a philosophical inquiry / Robin Brown and James Ladyman.
Description: Abingdon, Oxon ; New York : Routledge, 2019. | Includes
bibliographical references and index.
Identifiers: LCCN 2019000841| ISBN 9780367201333 (hardback : alk. paper) |
ISBN 9780367201340 (pbk. : alk. paper) | ISBN 9780429259739 (ebk.)
Subjects: LCSH: Materialism–History.
Classification: LCC B825 .B74 2019 | DDC 146/.3–dc23
LC record available at https://lccn.loc.gov/2019000841

ISBN: 978-0-367-20133-3 (hbk)
ISBN: 978-0-367-20134-0 (pbk)
ISBN: 978-0-429-25973-9 (ebk)

Typeset in Joanna and DIN
by Swales & Willis, Exeter, Devon, UK

Contents

Contents

The principal aims of this book are to provide the reader with an introduction to the history of the philosophical doctrine of materialism and to outline the elements of contemporary materialism, now known, for reasons explained in the text, as 'physicalism'. The references provided serve as a guide to further reading for those readers who wish to pursue both these areas in greater depth. The book has been written with the non-specialist reader in mind, but it is also intended to be of interest to those working in both philosophy and the history of ideas.

The chapters are of quite different kinds. Chapter 1 is a general introduction to the basic ideas at the heart of materialist philosophy. Relations to rival and kindred philosophical traditions are discussed.

Chapters 2, 3 and 4 cover the historical development of materialism from the first millennium BCE up to the conclusion of the nineteenth century CE. A short summary of such a huge topic is necessarily very selective, but the material chosen gives a general overview of the intellectual climate of the relevant period, and describes the place of materialist thought in that setting. An account is given of the work and influence of key philosophers who have a significant place in the history of materialism.

Chapters 5, 6 and 7 contain the core philosophical ideas and theses of contemporary materialism, and are the most demanding. They explain the concept of supervenience, which has a central place in contemporary physicalist thought, and discuss its far-reaching implications.

Chapter 8 is the concluding chapter. It considers the place of physicalism in the contemporary philosophical scene and in modern society at large. As may be evident, materialism is a philosophical doctrine that is not and cannot be confined to academic cloisters.

It may seem foolhardy to write a brief introductory book for the general reader in a field as contentious as philosophy. The authors are likely to be admonished for partiality, selectivity, over-simplification and subjective bias, and are likely to be guilty as charged, to a greater or lesser extent. Nonetheless, in the case of materialism it is important to undertake this task because, of all topics in epistemology and metaphysics, which together lie at the heart of philosophy, it is one of the most significant for people who otherwise have little or no interest in philosophy.

Ideas about what kind of stuff the world is made of have always been at the forefront of human thought, in some form or other, and few if any philosophical theories have aroused as much passion. The wars of religion in Europe that followed the Reformation may have had their origins in disputes about money-making by the Church, but denying the doctrine of the Trinity or that of Transubstantiation – both of which are purely metaphysical doctrines – became illegal. At different times and in different societies there has been extraordinary intolerance for some answers to ontological questions – those concerned with the issue of what sort of stuff exists. Even today there are several countries where it is a capital offence to have certain beliefs about the nature of the world, and about what kinds of things there are.

There are two interconnected strands to the history of materialism; there is the intellectual development and exposition of the philosophical claim, and there are the lives of materialist thinkers and philosophers who, while not materialists themselves, have played a key part in the development of materialism. This book discusses both. It is somewhat partisan in being admiring of the great thinkers in the materialist tradition, which includes one of the greatest poets of the ancient world, Lucretius. This admiration is partly for the ideas, but it is at least as much for the stance these thinkers have taken in the intellectual, social and political world they inhabited.

Prior to the twentieth century, materialist thinkers were in the vanguard of the cause of tolerance and free thinking. For reasons that will become clear, a certain kind of materialism became part of the foundational metaphysics of what has come to be known as the Radical Enlightenment, the great achievement of the Western intellectual tradition (Israel 2002). In the twentieth century everything changed; materialism became untethered from the Enlightenment tradition and, for the first time in history, regimes promoting a materialist ideology achieved state power. Materialism became associated with mass incarceration and murder. This shows, if further evidence were needed, that a theory about what the world is like has no necessary connection with the question of how human beings should behave.

We would like to thank Andrew Pyle and Jan Westerhoff for their very helpful comments on Part I of the book. We also thank the readers appointed by Routledge for their thoughtful and insightful reports. Finally, we would like to thank Tony Bruce at Routledge for his encouragement and enthusiasm in the process of bring this book to publication.

A preliminary disambiguation

In much contemporary discourse the word 'materialistic' refers to a way of life – 'excessive devotion to bodily wants or financial success', as the *Chambers English Dictionary* puts it. If the followers of such a way of life can, by extension, be classed as materialists, it is important to emphasise that this book is not about them or their *credo*. This book is about *philosophical* materialism, which at heart is a theory about the kind of things that exist. To adopt such a philosophical stance has no necessary connection with any particular attitude about how life should, or should not, be lived. Indeed, it is commonly, though not universally, agreed that an injunction to act in a certain way cannot be derived from a statement of how things are – an 'ought' cannot be derived from an 'is'.

While philosophical materialism may have no logical connection with any ethical system or way of life, asserting that only material things exist, and thereby denying the existence of spiritual things, does perhaps suggest that one should only be interested in material things, and seek one's rewards in life rather than in some afterlife. Hence, there are connections between philosophical materialism and what might be called 'hedonistic materialism', which is the view that life should be devoted to material pleasures.

However, errors arise from the ambiguous use of the single term 'materialism' for both. Furthermore, hedonistic materialism only degenerates into the way of life referred to in the dictionary definition quoted above – call it 'decadent materialism' – if a particular choice is made concerning which material things and pleasures to pursue amongst many possible ones. The natural world of flora and fauna, the arts and sciences and technology and engineering, to name just a few areas of human endeavour – all these may engage the interest of the philosophical materialist as much as, if not more than, fine food, fast cars and money. Nothing in philosophical materialism implies greed or gluttony.

The origins of philosophical materialism lie in the ancient world and arose in contrast to religious schools of thought (as Chapter 2 explains). It was a time when philosophical thought always had an ethical strand. The religious schools derived much of their teaching on the ethical life from their religious doctrines, including, typically, worship and rituals of sacrifice. In the later, monotheistic traditions, the glory of the spiritual stood in contrast to the 'lowly' pleasures of the body. With no religious belief to turn to as the bedrock of philosophical materialist ethics, materialist schools in both the East and West named the pursuit of pleasure as the true goal of life – but there are many different kinds of pleasure, as pointed out above. The most famous materialist of the ancient world, Epicurus, lived ascetically, along with the great majority of philosophically inclined people of his time in Athens, and taught, in his school, the 'Garden', that this was an appropriate way to live. Yet the great Roman poet Horace wrote of 'the sty of Epicurus', which is an outrageous calumny. Although critics can cite instances in his writings that are ambiguous on the question, for Epicurus the path to pleasure was by no means associated with excess or lavish taste.

Materialism was, until the twentieth century, associated with the liberal or radical traditions of the societies in which it occurred, for the straightforward reason that it stood in opposition to the prevailing conservative religious orthodoxies of the time. As such, materialism contrasts with more ascetic, self-denying styles of life that were based on religious doctrine, implying these styles of life are based on falsehoods and, therefore, largely pointless. Fasting, and other, sometimes more dramatic, self-inflicted physical torments, were rarely valued by philosophical materialists, but they did not promote decadent materialism as an alternative. They are accused of doing so because their enemies considered their actual views so dangerous. As a consequence, proponents of philosophical materialism have faced intolerance and persecution for long periods of time. That intolerance continues in many places today. Of course, materialists are not the only people to have faced persecution on account of their beliefs. Many religious people have suffered the same fate, and, since the turn of the twentieth century, that persecution has sometimes been, regrettably, at the hands of materialists.

An outline of the history of materialism

Part I

The heart of materialism

One

Introduction

Metaphysics is that branch of philosophy concerned with the most basic questions about reality. Ontology is that branch of metaphysics that is concerned with the question 'what exists?' Materialism is an ontological theory that presupposes an intuitive concept of *space*, and the primary claim of materialism is that the only things that exist are those that occupy space. In the Latin of medieval philosophy, these are *res extensa*, extended things. Clearly there is a negative implication of materialism. The existence of spirits, ghosts and, crucially, transcendent beings such as the god of the monotheistic religions is denied by materialism. The *res cogitans* of medieval philosophy, thinking substance, according to materialists does not exist.

The problem is that this kind of materialism seems to be false. Undoubtedly there are things that do not occupy space at all, but the existence of which we would not seriously question. Candidates for such things include thoughts, velocity and danger. Materialism should be revised to assert that what exists, in addition to *res extensa*, are things that *depend* for their existence on things that occupy space. In other words, without material things there would be no thought, velocity, danger or anything else. The positive content of materialism becomes a claim about some kind of dependence, and hence materialists need arguments to demonstrate that some given class of things depends wholly for their being on material things. It proves to be easier to demonstrate the dependency of some classes of things than others. For example, *velocity* is not a thing that occupies space, but it is relatively easy to show that things like velocity, while not in *themselves* occupying space, are dependent in the way required; without there being things that occupy space and that move, there would not be velocity.

Abstract entities, such as numbers, are more problematic. It is not feasible to think of the number two, for example, occupying space, even if any example of a sign for that number – for example, '2' – does occupy space. Statements like 'there are infinitely many prime numbers', which is a truth of arithmetic, seem to be ontological assertions. While materialists struggle with this kind of challenge, it doesn't seem to bother them unduly. There are various responses; many simply deny that the number two has any genuine existence at all, arguing that the whole edifice of arithmetic is an abstraction arising from the perception of collections of individuals that occupy space. Others argue that whatever kind of existence numbers have, it is irrelevant to the kind of ontological problems in which materialists are interested.

The most contentious subject matter for materialism is always psychological phenomena, in particular conscious phenomena, such as perceptions, feelings and thoughts, and, critically, free will and practical reason. It is at this point that the philosophical dispute between materialists and their critics turns from ontological to moral – and even political – concerns. For example, while anti-materialists sometimes argue that materialist philosophy promotes a cold disdain for ethical commitment, materialists argue that belief in life after death discourages people from demanding enough from the one life they surely have.

Metaphysical theories do not stand independent of epistemological theories. Epistemology is that branch of philosophy that concerns itself with knowledge and belief. Anyone asserting an ontological theory needs to have something to say about epistemology, if she is to be taken seriously, to answer the challenges – how do you *know* that what you say exists does, in fact, exist? Or, on what grounds do you *believe* that what you say exists, exists? Materialism is an ontological theory that is intimately connected to a particular epistemological perspective.

Aristotle's *Metaphysics* famously begins with the statement 'All men by nature desire to know'. Metaphysical theories and epistemological theories go hand-in-hand in the human project of satisfying that desire to know, and to understand the world that human beings inhabit. But it is prudent to add to Aristotle's statement that men by nature want to *feel* that they know. People, or at least those Aristotle is talking about, don't like the feeling of not knowing; it makes them anxious and uneasy, and in the middle of a violent and as-yet-unexplained thunderstorm, frightened. Some theory about what is going on, and preferably a readily understood theory, eases some of the anxiety not-knowing brings. The issue of whether or not that theory is *true*, whether or not it is *genuine knowledge*, is not of primary importance in stilling disquiet in the mind.

No sooner have we begun our inquiry in ontology than we have been obliged to consider epistemology, and then straightaway we must consider psychology. Faced with the 'desire to know' – we could call Aristotle's idea the *epistemological urge* – we face the question of what methods to employ to achieve knowledge, and – here's the psychology – to satisfy us and make us believe we have found knowledge.

The people of many of the first human societies, though apparently not all, developed theories that provided an account of the origins and nature of the world and natural phenomena. Two features of these theories are important: first, that such theories have an important role in strengthening social cohesion – a society can feel more cohesive if its members share a common outlook. The second property is that these theories very often involve reference to gods and spirits in the accounts given of natural phenomena.

Together, a society's ontological perspectives expressed in these theories can be understood as the society's *worldview*. The epistemology at the origins of the worldview is often hidden. Consider a society in which someone thought up the idea that thunder was the expression of a powerful being's anger. To the modern understanding this is a projection of human emotional experience onto the world; through the link of loud noise and violent effects, thunder becomes associated with anger. Once such a worldview becomes established, for subsequent generations the source of belief becomes an authority that provides a canonical interpretation of events in terms of the supernatural agent's temperament. The authority can be people – the elders or the priests, for example – or it can be, additionally, in societies that developed written language, a text, sometimes a holy book.

From the earliest times, materialists and their critics have been in dispute in a way that can be understood as a dispute between alternative worldviews – that is, a dispute involving divergent ontological and epistemological claims. A dispute about such matters can generate a lot of heat, as we shall see. But there is a critical asymmetry between the materialists and their opponents. In virtually all instances before the twentieth century, whenever there was a more-or-less established worldview, materialism was in the opposition to it. There were periods in the ancient world when there was genuine freedom of thought, and materialism was free to argue its case with alternative ontological theories, but, for much of the time, proponents of materialism were considered to be a kind of dissident, or outsider, and as such were susceptible to, or threatened with, intolerance and persecution. Thinking the 'wrong' way about the nature of the universe could, and often did, prove to be dangerous.

Before describing the history of materialism, a little more needs to be said about the alternative perspectives that developed in epistemology. What might be the possible sources of knowledge? Traditionally, in philosophy, two contrasting answers to this question are *empiricism* and *rationalism*. At its crudest, empiricism finds the source of all knowledge in our perception of the world through our sense organs, while rationalism names the source of our knowledge as our reason. It is not only to someone unfamiliar with the philosophical tradition that this dichotomy may seem odd, if not downright absurd. Human beings cannot really be imagined without sense organs, and perceptions would seem to be necessary to provide at least some of what reason takes as subject matter. Equally, sensory perception without some application of reason is going to provide nothing at all beyond sense perceptions, which cannot be, in and of themselves, knowledge. Knowledge can only come from an interplay of reason and perception.

Other problems with empiricism concern the analysis of just what it is that can be taken to be the sensory input to our sense organs from the external world. A common-sense view would take it that we see and touch, for example, a table. But the more sceptical empiricist may insist that this is just a hypothetical construction from the raw data of shapes of colour and tone that the eyes perceive, and the 'feel' in our fingers.

The true nature of the contrast between empiricism and rationalism lies in the critique of our reasoning alongside the critique of the evidence gained from perception. Perhaps the most striking example of the dispute concerns Parmenides, the pre-Socratic Greek philosopher who argued that motion was impossible. It would seem evident that there is motion, from what we perceive of the world, but Parmenides believed he had sound arguments to prove that motion was impossible, and therefore the evidence of the senses was unreliable. Parmenides needed to perceive *apparent* motion before developing the argument for motion's illusory ontological status. The empiricist cannot rely simply on that perceptual evidence; she has also to discredit the reasoning that led to Parmenides' claim.

Materialist epistemology has evolved to coincide with what can be identified as the epistemology of modern science. With this perspective, empirical evidence is a necessary but not sufficient element for knowledge. Any assertion about the nature of reality may start from reasoned speculation, but must be subjected to test and critique, and, therefore, faces potential rejection. Theories that cannot be tested rule themselves out from scientific credibility. Finally, the scientific stance never relinquishes some element of provisionality, of tentativeness, in the details of theories that are espoused.

The heart of materialism is the withholding of belief in the existence of certain kinds of entity and certain kinds of phenomena. There are, it claims, no gods and devils, no ghosts, no spirits. There is, also, no such thing as Providence, or Luck or Fate. Reality consists of material things and things that are wholly dependent for their existence on material things. Their existence is controlled by laws of nature that are independent of Will. The development of the world is not directed by any pre-established plan. There is no predetermined End, good or bad, to which change is directed.

Materialism believes the vital psychological phenomena of our human and animal existence are wholly dependent on the material nature of our bodies. Though it remains obscure to human understanding, they emerge, in some way or other, from our material being. There is no soul independent of our bodies, let alone one that could survive the destruction of our bodies. There is no afterlife. A human life is a temporary phenomenon, normally encompassing a timespan of less than 100 years.

Materialism has humility in its heart, though it is admittedly sometimes hidden. It claims no path to knowledge other than through scientific endeavour. It holds no conviction that human beings can reach a true Theory of Everything, but it equally presupposes no set limits on human knowledge. It knows there are vast areas of reality about whose workings we know little or nothing, but eschews the adoption of scientifically inadequate theories to satisfy our quest for epistemological peace of mind. For sensible materialists, psychology in general, and consciousness in particular, remain a mystery. The optimists believe the mystery will be resolved; the pessimists are not so sure.

Materialists deny any objective grounding for morality and the notions of Good. Typically, morality is viewed by materialists as the codified rules that facilitate social stability. Such a code may be subject to criticism of various kinds. It may be accused of hypocrisy, if its proponents claim it serves the entire society while its critics see its purpose as maintaining the power of social elites. Alternatively, it may be shown to contradict principles, such as fairness, that the society endorses. But materialism as such can only offer a critique to the suggestion that the code is grounded in an objective legitimacy, the source of which is often identified as a supernatural figure or an authoritative sacred text.

That said, it can be argued that, paradoxically, there is an ethical perspective, if not actually at the heart of materialism, then at its side as a close companion throughout its history, until the twentieth century. Evidently, this perspective does not derive from the core philosophy of materialism but rather from the social experience of its proponents. As mentioned above, until the twentieth century materialists were commonly outcasts in society. In some periods, materialism found itself in a free-thinking milieu where it could flourish alongside rival ontological and epistemological perspectives, but for the greater part materialism has been on the margins of society, disapproved of, barely tolerated. Its adherents were seen as outsiders, opponents of established norms. They were not 'right-thinking', and, in consequence, they were commonly mocked, derided, vilified – and persecuted. It is suggested in Chapter 4 that the apogee of materialism is in the eighteenth century, and here we see it clearly adopting an ethical stance of freedom of thought. So it can be said that materialism's ethical companion is toleration. Until the twentieth century, materialists were typically advocates of the right to divergent opinion, and of opposition to the imposition, by authorities, of beliefs and ways of thinking.

Accordingly, materialist epistemology requires that the outline of the heart of materialism is not read as a statement of dogma. The materialist asks for evidence for any statement about the nature of the world, but it should not, in the true scientific spirit, claim certainty about anything. All scientific theories are held with a degree of caution, and insofar as the belief that there is no god is a theory about the nature of the world, the materialist acknowledges the possibility that his or her belief is false. Materialists believe that there is neither credible evidence nor powerful argument for the existence of god, so that there is a negligible probability of theism being a true theory.

As stated above there is no necessary link between freedom of thought and materialist ontology and epistemology, and by the time materialists gained state power in some countries in the twentieth century the link with such ethics had been shattered. This was to devastating effect – not only for the victims of persecution at the hands of states governed by materialists, but also because it changed the standing of materialism in the intellectual realm. Although always disapproved of by the religious, materialism had previously been in the camp of tolerance and free thinking. In the space of fifty years its social standing was diminished by association with much darker social trends rather than enlightenment.

As the course of the development of materialism is discussed in what follows, it is shown that contemporary materialism, under the name of physicalism, has adopted some far-reaching modifications of materialism. However, the heart of physicalism is shown to be both the true heir to, and a natural extension of, the heart of materialism.

Materialism in dispute with other ontologies

The critique of materialism has two main strands – one is that it is false and the other that it is not only false but also dangerous. For present purposes the second critique can, for the time being, be set aside. The first threatens the materialist with disproof and theoretical rejection. The second threatens the materialist with persecution. While the second is likely to concentrate the mind more than the first, it is the theoretical objection that can be, and has to be, addressed with reason.

It makes sense to start, and quickly dispense with, a line of attack that both materialists and their critics engage in – it is to declare that the position of the opponent is *absurd*. Opponents of materialism argue that materialism is absurd – how can mere matter produce psychological phenomena? If all there is, fundamentally, is matter in space, how could consciousness possibly appear? How could good and bad, right and wrong, have any meaning? If everything boils down to matter driven blindly in its motion by the laws of nature, how could a human being possibly have free will?

Materialists have their own version of this non-argument. Stories of gods throwing hammers making crashing sounds in the sky, or driving chariots through the sky to give us daylight, are the pinnacle of absurdity, if supposed to be taken literally. And so stories of heaven and hell, of judgement and punishment and reward after death, are equally just so much nonsense.

It is sensible to dispense with these accusations because they are not serious arguments. The appeal to a notion of absurdity can, generously, be taken as an appeal to intuition as encountered often in philosophy. Obviously, people have different intuitions. It's one thing to convince ourselves of the correctness of a position because it feels intuitively correct to us, but we need do better than that to convince someone who doesn't share our intuitions.

The central criticism that materialists put forward to challenge non-materialist ontologies can be stated in a more sophisticated way. It is that the theories they oppose lack genuine evidence. The materialist resists the argument from authority as a valid ground for belief. More particularly, the materialist demands some combination of rational argument and empirical experience as a necessary condition for justified belief. They believe the non-materialist fails to provide convincing arguments and perceptual experiences that can be considered as genuinely evidential. On the other hand, given that matter and psychological phenomena *seem* to be different kinds of thing, it falls to the materialist to provide some account of how the psychological – and the ethical – arise from a wholly materialist world.

In this way, in the materialist stance there is both defence and attack, and there is not a unified materialist response to criticism. Some materialists are more sure, more convinced, more belligerent, than others. Materialists hold different conceptions of the phenomena under question – some, for example, bizarrely, even deny the existence of psychological phenomena, thereby supposedly eliminating the problem with a stroke of the pen.

More plausibly, materialists have produced arguments to show *free will* to be a phenomenon radically unlike the pervasive but primitive idea of it. It looks as though a 'me' stands outside the physical order and decides what course the future is going to take – it is down to this 'me' and nothing else whether the window is opened or remains shut in the next thirty seconds. Materialism demands a radical and detailed analysis of what the 'me' in question is like.

A frank materialism acknowledges major gaps in our understanding with this two-pronged defence; gaps in knowledge should not be filled with theories that lack scientific credibility, and gaps in knowledge are in general going to be filled by science, and not philosophy. Philosophy is the servant of science, albeit an essential one, and not its master. There is a critical corollary here – *science could refute materialism*, by discovering non-material phenomena. As explained in Part II, something like this is in fact what has happened, prompting the need for materialism to evolve into physicalism.

Materialism can also be seen to offer an account of non-materialist theories, and of how they appeal. It can, from one perspective, be seen as expressing a psychoanalytic critique, centuries before Freud; animists and religious people, it seems to be claiming, are simply *projecting* their own psychological concerns onto the world. Good standing, as a child, with the elders, and as an adult with your fellow men and women, are concerns that are imagined lived out in your dealings with the natural world. But the earth and the sky, the thunder and storms, the earthquakes and volcanoes, have no interest in you. They don't have interests in anything, because they don't have interests at all. Parents can be pleased with their children when they are well-behaved and angry with them when they are badly behaved, but there is nothing in nature that is pleased with you when the harvest succeeds, or cross with you when it fails. Parents might be pleased with the child when he or she forgoes something desired in order to appease them, but there are no gods pleased with you because you have killed a sheep, or a young virgin, in their honour. Religious perspectives, these materialists would claim, are *infantile*.

The relation of materialism to allied traditions

It is helpful to name some close relations and to note what distinguishes them from materialism.

Perhaps the closest next-of-kin to materialism is *atheism*. Also an ontological theory, atheism makes the wholly negative claim that there is no god or gods. It is evident that a definition of 'god' is required before atheism can be expressed coherently, something that is not a requirement of materialism. If god were potentially material, in the sense of occupying space, then for materialists it would be an open question whether or not there was a god. There would be the same principled rejection of the claim that something non-material exists.

The idea of 'god' has, of course, itself undergone profound changes in the Western tradition. It is not that clear what the ontological status of the gods of the *Iliad* and the *Odyssey* is. Did they occupy space? They lived on a mountain, and sometimes took the form of people and animals, so perhaps they did. On the other hand, they were immortal, so what was taking up space wasn't like the flesh and blood of human beings. By the time we reach the age of the great monotheisms, Judaism, Christianity and Islam, the god being dealt with is specifically denied by materialism, because the god is specifically identified as immaterial, and the idea that the god of these monotheisms is material was considered a heresy. Therefore, in this period, all materialists were atheists, but perhaps not all atheists were materialists. An obvious example of a non-materialist atheism would be the Buddhist tradition, which holds that there is no god but which does believe in the transmigration of souls. Materialism, and non-materialist atheism, in the Eastern tradition, is discussed in Chapter 2. To the religious, both materialism and atheism are considered – with good reason – forms of *scepticism*. Both promote doubt about the supposed foundations of religious belief, and deny religion's claim to knowledge, based as it commonly is on individual revelation, dubious reasoning and people or books granted supreme authority.

As indicated above, *agnosticism* may, formally, be a closer relation to materialism than atheism, insofar as the materialist avoids dogmatic statements, but, the 'not-knowing whether there is a god' of the materialist agnostic is not so very far from the 'knowing there is not a god' of the atheist insofar as for either view there is no reason to take part in religious practice.

Perhaps the most important perspective, when considering the social impact of materialism as an ontological theory, is simply to recognise that it has, as a consequence, the denial of a non-material deity. But an interesting additional question comes to mind; it was noted above that materialism is obliged to recognise the existence of some non-material things, and does just so long as those non-material things are wholly dependent for their existence on material things. What, then, if the deity were dependent for its existence on the material world? Of course, in the established traditions of the world's major religions, the very idea is an outrage. However, it does seem that there is a significant number of people who, while they find it hard to swallow the ontological claims of the traditional teachings, want to retain a religious sensibility in their lives. This can involve not only a sense of the spiritual, which may or may not involve engagement in ritual, but also an idea of something that would naturally go by the name 'God'. From the 1960s onwards, in the more liberal strands of Protestant Christianity, there has been a culture of uncertainty about just what the ontological claims of the teachings are. It is not uncommon to hear people talking of finding God within themselves, or of God being manifest in good deeds. Sometimes God appears to be imagined as something like an idea. The materialist claim that our human psychology is wholly dependent on our material selves can accommodate a god that is essentially a human idea.

In any case, there are good reasons for distinguishing the tasks of defending materialism from the tasks of promoting atheism. The atheist is of necessity involved in a confrontation with theists, and there is much disputation here that the materialist can reasonably bypass. Consider the following remark by Rupert Shortt, the religion editor of the *Times Literary Supplement*, in a book review in that periodical.

> Informed Jews, Christians and Muslims standing at different points in the same field would insist that God is not a thing who competes for space with creatures. You cannot (to posit a crazy thought experiment) add up everything in the universe, reach a total n, then conclude that the final total is n + 1 because you're also a theist.

God belongs to no genus; divinity and humanity are too different to be opposites. By definition, then, no physical analogy will describe our putative creator adequately. We are migrating off the semantic map. But light is amongst the more helpful. The light in which we see is not one of the objects seen, because we apprehend light only inasmuch as it is reflected off opaque objects. From a monotheistic standpoint, it is the same with the divine light. The light which is God, writes the philosopher Denys Turner, we can see only in the creatures that reflect it. Therefore ... when we turn our minds away from the visible objects of creation to God, ... the source of their visibility, it is as if we see nothing. The world shines with the divine light. But the light which causes it to shine is itself like a profound darkness.

(TLS, 16/12/16, p. 4)

Who knows what proportion of Jews, Christians and Muslims worldwide are informed in Shortt's terms, and what proportion is of the n + 1 school. Bowing to Shortt's authority on these matters, the suggestion here is that materialists are excused from this dispute as they are essentially concerned with the n things and those things wholly dependent on them for their existence.

Materialism also has close connections with *humanism*. Humanism gets its name from a denial of the superhuman deities of religion, and also involves an ethical perspective associated with the principles of the Enlightenment, to be discussed in Chapter 3. While not necessarily being committed to materialism, most humanists adopt a materialist ontology.

Materialism, atheism and humanism are related to naturalism. This is essentially an epistemological doctrine that rejects any but natural explanations of natural phenomena – explanations, in other words, that eschew concepts like Providence, Divine Intervention, Fate and other agents of a supernatural kind. In *The Naturalistic Tradition in Indian Thought*, Riepe identified the following six elements of naturalistic thinking.

1 The naturalist accepts sense experience as the most important avenue of knowledge.
2 The naturalist believes that knowledge is not esoteric, innate, or intuitive (mystical).
3 The naturalist believes that the external world, of which man is an integral part, is objective and hence not 'his idea' but an existent apart from his, your, or anyone's consciousness.
4 The naturalist believes that the world manifests order and regularity and that, contrary to some opinion, this does not exclude human responsibility. This order cannot be changed merely by thought, magic, sacrifice, or prayer, but requires actual manipulation of the external world in some physical way.
5 The naturalist rejects supernatural teleology. The direction of the world is created by the world itself.
6 The naturalist is humanistic. Man is not simply a mirror of deity or the absolute but a biological existent whose goal it is to do what is proper to man. What is proper to man is discovered in a naturalistic context by the moral philosopher.

(Riepe, 1964, pp. 6–7)

There is clearly room here for perspectives that are neither materialist nor atheist, but it is equally evident that materialism and atheism are members of the broader family of naturalism. Perhaps 'naturalistic materialist humanism' would be the preferred, if overblown, name for the perspective of many theorists seeking not only ontological and epistemological theories, but also an ethical outlook.

It is time to turn to the beginnings of materialist thought in the ancient world.

Two

Introduction

There is a widespread belief that materialism is a modern idea, and a Western idea. Neither is true. The roots of materialism are ancient and it is well represented in the East. There were important sceptical traditions in many ancient societies where there were dominant religious traditions. In the West, following the decline of Rome, Christian teaching dominated metaphysical thought. Over time some of the teachings of the great non-materialist thinkers of classical Greece, Plato and Aristotle, were incorporated into Christian philosophy and doctrine, while the ancient atomists were largely forgotten. In India and China too religious ideologies came to dominate philosophical thought.

It is a mistake to think that the Eastern and Western philosophical traditions are at odds with one another, or that there were few points of contact between them. There is compelling evidence that there was interplay between the Greek world and India in the final 500 years BCE, largely through the geographical and social bridge afforded by the ancient Persian Empire, which linked the Mediterranean world and the sub-continent. Furthermore, the idea that Eastern thought is fundamentally mystical and Western thought is fundamentally hard-headed and the root of modern science is just plain wrong. Atomism and atheism were both discussed in ancient India and China, and science owes much to places such as Alexandria and Baghdad, which can hardly be described as 'Western'.

There are a number of problems with knowledge of thought in the ancient world. First, the primary sources available are scant. Things were written down on materials that over time perished. There was no printing, so over the centuries we have a great debt to armies of scribes who painstakingly copied texts so that their teachings would survive. Later in this chapter an extraordinary strand in the history of materialism is mentioned in which the most important work on materialism in the ancient world was preserved thanks to the work of scribes about whom we know nothing except that they were in all likelihood hostile to materialism.

There is another reason of a quite different kind why there is a paucity of works of some authors and not others. The dawn of the Common Era saw a huge rise in the *suppression* of thought as detailed in this chapter. Teachings that were considered erroneous came also to be seen as dangerous, and therefore subject to criticism and actual destruction.

Contemporary knowledge of a philosopher, whose work survives only in odd fragments, is largely based on the reports of another, later philosopher or philosophers or other writers such as playwrights. There are occasions when materialist thinkers are known by accounts of their views by opponents whose intention was to discredit and disprove the materialist teaching. The reliability of their reports is obviously suspect.

There are also contemporary attitudes that have arisen from developments in human thought that are remarkably recent, and which need to be kept in mind when considering the thought of early civilisations. Written language has a history of no more than 5000 years. There is little doubt that speculative thought long predates this, but systematic theorising, and the dissemination of ideas beyond a relatively small locality, would have depended on text. Many of our ways of thinking arise from developments in the last 500 years. Most obviously, the clear distinctions made now between philosophy, religion and science were not present in the epoch that concerns us here – roughly the last 500 years before the Common Era. The ideas associated with the scientific method – observation, experiment, evidence and confirmation and falsification were not widely accepted. Indeed, in the ancient world, these ideas were understood only in a very primitive way, and on the whole the technologies and measurement techniques needed to accompany them were unavailable. A year's study of logic now provides a student with much more knowledge of logic than was available to Aristotle, the greatest logician of the Hellenic period. Similarly, a year's study of chemistry now provides a student with much more knowledge of chemistry than was available to anyone before 1600 CE. It is startling that the ancient ontological categories – earth, air, fire and water – had a fundamental role in the natural sciences until the seventeenth century, along with other theories, such as those of the four humours, that seem quaintly ridiculous. (The theory of the four elements can be seen more charitably as an early version of the distinction between the different states of matter – solid, liquid, gas and plasma.)

This chapter is about materialist thought. It is not an account of thinkers who would think of themselves as materialists, or that are appropriately identified as materialists in a modern sense. It is rather that in their writings there are specifically materialist perspectives on the nature of reality. In fact, elements of materialist thinking are present in many of the early philosophers, but in those discussed here the materialist elements are most pronounced and have had the most influence on later materialist thinkers.

Similarly, it can be difficult to separate out a specifically materialist element to strands of thought that are primarily self-identified as sceptical or critical or in opposition to some pre-existing or established set of beliefs. In studying a cultural climate where science, religion and philosophy coalesce, it would be an unnecessary and probably futile task to try to disentangle specifically atheistic thought from purely materialistic thought.

The first port of call is, in many ways, the ancient school closest in important respects to the fully articulated materialism that takes shape more than 2000 years later. It may be surprising to many readers that the most avowedly materialist school of the ancient world is to be found in India.

Materialism in the Indian tradition

The intellectual tradition of India in which philosophy and religion are commonly intertwined can be divided broadly into two traditions. The beginnings are the teachings of the Vedas, dating from the middle of the second millennium BCE and culminating in the Upanishads of the sixth century

BCE. The orthodox pro-Vedic (Brahminical) schools that developed preach the existence of an afterlife, the soul and the transmigration of souls. Many but not all are also monotheistic. Schools by the name unorthodox, or heterodox, or *nastika*, reject to some degree the Vedic teaching or at least dispute the orthodox interpretation of it. While much heterodox teaching denies the existence of God, the principal heterodox schools, Buddhism and Jainism, accept the existence of spiritual entities such as would be associated with the transmigration of souls and reincarnation.

The school of concern here can be viewed as a heterodox school or, more appropriately, as standing between and apart from the two traditions. It is known as the Lokayata, or Carvaka, and it is openly, flagrantly, materialistic.

Joshi writes

> In Indian philosophy the systems which are generally regarded as atheistic are the Carvaka, the Samkhya, the Mimamsa, Buddhism, and Jainism. The term atheism, when applied to a system of thought, usually means that the system has no use for the concept of God and that it is opposed to all forms of spiritualism and religion. Judged in this light, the Carvaka is the only true form of atheism. The Samkhya, the Mimamsa, Buddhism, and Jainism are atheistic systems with a difference, for while they deny the reality of a personal God, they openly embrace spiritual and even religious ideas.
>
> (Joshi, 1966, p. 189)

The tradition, then, embodies a teaching that Joshi calls a true form of atheism, and which is, evidently, materialist.

There seem to have been distinct strands in the epistemological perspectives of the Lokayata tradition. While there was a general acceptance of the empiricist view that the basic source of knowledge is sense perception, there are differing opinions about the status of knowledge derived from different kinds of inference. There may have been extremely sceptical positions in the tradition, but also more moderate positions that accepted that inference may sometimes yield genuine knowledge. These differences mirror controversies in the empiricist tradition in British philosophy, arising from the work of Locke and Hume.

In the *Oxford Handbook of Atheism*, Frazier (2013) writes

> In its most unambiguous form, atheism developed as a distinct and well-formed philosophical school of thought referred to from the sixth century BCE at least into the medieval periods, known as the Lokayatas or 'worldly ones', which propounded the material nature of the world and the non-existence of the soul. As there is no further or higher reality from which to take our ethical cue, happiness (understood in terms of pleasure – *kama* – the fulfilment of our desires) in this world is the only self-evident good, to which our efforts should be directed.
>
> (p. 370)

Frazier describes a culture in which there is a broad range of metaphysical perspectives that are able to challenge and confront each other. At the same time, she suggests that the materialists did attract growing disapproval from their religious opponents, and speculates that one form of their persecution may have been the destruction of their writings, of which only fragments now remain.

The Lokayatas are a recognised school of thought in the medieval period, and in a compendium of such schools compiled by Madhava there is a suggestion that they had a considerable following.

> The mass of men, in accordance with the manuals of politics and enjoyment, considering wealth and desire the only goals of humanity, and denying the existence of any object belonging to a future world, are found to follow only the doctrine of the Carvakas.
>
> (Madhava, 1978, p. 2, quoted by Frazier, 2013, p. 371)

Frazier, summarising Madhava's account, goes on

> ... the Lokayatas [believe] the elements of air, earth, fire and water are the sole constituents of reality, from which all things (including consciousness) are ultimately derived. This view is grounded in a firmly epistemological starting point: Lokayatas are Humean empiricists who believe that perception is the only ... valid source of knowledge.
>
> (pp. 370–1)

Madhava depicts the Lokayatas as hedonists (an early example of the coincidence of the two different conceptions of materialism discussed in the preliminary disambiguation). It remains unclear to what extent the pursuit of happiness was understood shallowly by the Lokayatas, as opposed to this being a slur on them by their religious opponents. It does

seem clear the Lokayatas had little time for the ascetic life –
Frazer quotes from a Lokayata text cited by Madhava: 'If
anyone were so timid as to give up a visible pleasure, he
would indeed be as foolish as a beast'.

Frazier goes on to describe Madhava as depicting the
Lokayatas

> ... as what we today call epiphenomenalists: nothing
> exists but the material elements, and thus the phenom-
> enon of consciousness can be attributed merely to
> a particularly bodily combination of these elements
> which generates thought ... the Lokayatas liberated them-
> selves from the moral narrative of karma and rebirth ...
>
> (pp. 371–2)

With regard to religion and the religious authorities,

> The Lokayatas also posed a critique of schools that
> claimed normative authority via scripture and divine
> mandate. Madhava describes them as ridiculing the sac-
> rifices of priests, and dismissing the Vedic rituals as
> merely a source of income, otherwise to be considered
> a waste of time and energy. They also reject the scrip-
> tures as untrue, self-contradictory and tautologous, and
> leading to contradictions. They deny any future exist-
> ence, and claim the closest thing to a supreme being is
> the earthly monarch ... People are advised to give their
> religious donations to living people in need.
>
> (p. 372)

There are further interesting remarks by Karel Werner
(1997) in his chapter on non-orthodox Indian philosophies

in *The Companion Encyclopedia of Asian Philosophy*. Commenting on the criticism of the Lokayatas as hedonists, he writes,

> Although the hedonistic aspect of the Lokayata ethics was often overemphasized in the preserved accounts which come invariably from opponents, and granted that there must have been some realistic grounds for the exaggeration, it is nevertheless also clear that, as in the case of the Greek equivalent of Lokayata, the philosophy of Epicurus, there were also positive aspects to Lokayata. There is some evidence that the intellectual pleasures were also prized and that the pursuit of sensory pleasures was incompatible for many with perceiving, let alone causing, suffering to others, especially by killing. Hence a further reason for Lokayata condemnation of animal sacrifices. Some Lokayatists seem even to have condemned war for the same reason.
>
> As the preoccupation with refuting Lokayata philosophy in orthodox and other philosophical writings in India lasted for several centuries, it has to be assumed that it must have had a significant following during that time and that it must have reached a considerable degree of theoretical elaboration, especially in the field of logical argument.
>
> (pp. 119–20)

For a recent, book-length treatment of the Lokayata tradition, see Gokhale (2015).

It is striking how modern these ancient thinkers can sound. While the science is primitive, and the logic and philosophy can seem limited, on the basis of the scant evidence available to us, nevertheless the Lokayata seem to have embraced the essentials of the materialist perspective outlined in Chapter 1. Moreover, while they lived for centuries in a relatively free-thinking intellectual environment, they often suffered the same disparagement as their later Western fellow materialists.

Atomism: the materialism of the Greeks

The story of materialism in the Western tradition begins with the philosopher Democritus, who was a younger contemporary of Socrates (but still often called a 'Pre-Socratic philosopher'). He is associated with Anaximander and Leucippus, who is reputed by some to have been his teacher. Democritus was born around 470 BCE, so materialist thought may have emerged earlier in India than in Greece.

Socrates, Plato and Aristotle deserve their eminence but other Greek philosophers sometimes seem insignificant in the company of such giants. This is because extremely little of their work survives. The principal reason there is so much of the work of Plato and Aristotle is that the Christian Church sought to incorporate their philosophies in canonical doctrine, and so preserved their writings – through the production of copies – with diligence. It is true the monastic scribes also copied the works of other 'pagan' authors, but the fate of their works was often to have their manuscripts rot – or worse, be actively destroyed. This was particularly true in the first centuries of the Common Era. Aristotle is the source of much of our knowledge of the early philosophers, because he wrote on their work and criticised it, providing a historical setting in which to propound his 'superior' philosophy. (Perhaps the other most

important ancient guide to the Greek philosophers is Diogenes Laertius (2015), who wrote in the third century CE.) In
particular, Aristotle admired Democritus, who was a prolific
writer, producing dozens of treatises on a wide range of
subjects. Although more survives of Democritus's work than
that of any other philosopher of the era besides Plato and
Aristotle, there are only fragments.

Democritus' fundamental thesis is that matter is not infinitely divisible. Rather, he believed there would be a finite conclusion to the process of dividing up a piece of matter, at
which point there would be many tiny indivisible bodies, too
small to be visible to humans. Democritus gave the name
'atom' to these tiny entities, *atomos* being the Greek word for
'indivisible'. He believed further in an infinite void –
vacuum – and held that atoms are in constant motion in it.

Democritus also thought there were infinitely many
atoms and that they come in infinitely many varieties, with
countless different shapes and sizes. The objects of our
world are complexes of atoms brought together by random
collisions, and the difference in their constituent atoms
accounts for the observable difference amongst the objects.
The only ultimate realities are atoms and the void. Democritus also thought that everything happens necessarily due
to the motions of atoms.

The philosophy is thus materialistic and atheistic, and
although Democritus believed in the human soul, it too
was composed of atoms – albeit special, spherical ones.

In *The Presocratic Philosophers*, Kirk and Raven (1964) have
written

> Atomism is in many ways the crown of Greek philosoph-
> ical achievement before Plato ... It was in essence a new

conception, one which was widely and skilfully applied by Democritus, and which through Epicurus and Lucretius was to play an important part in Greek thought even after Plato and Aristotle. It also, of course, eventually gave a stimulus to the development of modern atomic theory – the real nature and motives of which, however, are utterly distinct.

(p. 426)

It is questionable exactly how much current physics owes to Democritus, given that atoms as we know them are divisible, but the overall point made in this quotation is important. Recalling that philosophy, religion and science are not separate at this stage in history, Democritus' project is a very different one from that of the modern scientist. However, both he and the modern-day scientist seek empirical knowledge about, and naturalistic explanations of, the external world. Like Aristotle, Democritus was interested in the investigation of nature based on observation. Democritus, in his epistemology, thought the senses lowly as a source of knowledge, producing only 'bastard' knowledge, being subjective and being required to be processed by inductive reason. This seems to be a reasonable response to the problem of understanding the process beginning with sense perception and ending with knowledge, a problem for philosophy and science that has been addressed throughout the ages from the Lokayata on and up to the present day. Indeed, when Democritus argued that properties such as sweetness and odour are not amongst the properties of atoms but are appearances produced by quite different properties of atoms, he introduced a distinction that would be taken up by Galileo and become central to scientific thought.

In character, Democritus was cheerful. He was known affectionately as the laughing philosopher, although some called him the mocker – he may have been laughing at the foolishness of others. His ethics presage Epicurus, again wrongly seen as hedonistic, but also believing in the goal of cheerfulness or wellbeing, living a life of untroubled enjoyment.

Epicurus

The vituperative and groundless *ad hominem* attacks on the great materialist thinkers are a recurrent theme in this book. As the principal theoretical challenge to theistic doctrine and religious practice, materialists are seen not only as philosophical opponents but as ethically degenerate in various ways. If attacks by enemies are a measure of a materialist philosopher's significance, Epicurus is very important indeed. He is said to have written 300 books, but all that is left are some maxims and three letters that summarise his teachings on the philosophy of nature and on morality. The letters and one set of maxims can be found in Diogenes Laertius, who devotes the tenth – and last – book of his *Lives of the Eminent Philosophers* to Epicurus. He gives more pages to him than to any other philosopher, including Plato, the sole subject of Book Three, Aristotle and Socrates. He begins the account after some biographical details with a list of authors who have bitterly attacked Epicurus and accused him of all kinds of base behaviour. And then, at paragraphs nine and ten, he says

> 9. But these people are stark mad. For our philosopher has abundance of witnesses to attest his unsurpassed goodwill to all men – his native land, which honoured

him with statues in bronze; his friends, so many in number that they could hardly be counted by whole cities, and indeed all who knew him, held fast as they were by the siren-charms of his doctrine ... ; the School itself which, while nearly all the others have died out, continues for ever without interruption through numberless reigns of one scholar after another; 10. His gratitude to his parents, his generosity to his brothers, his gentleness to his servants, as evidenced by the terms of his will and by the fact that they were members of the School ... ; and in general, his benevolence to all mankind ...

Such extreme admiration raises doubt in the suspicious mind, and the following words come as something of a shock: 'His piety towards the gods and his affection for his country no words can describe' (Diogenes Laertius 2015).

In *Ancient Philosophy*, Kenny (2004) explains Epicurus' attitude to the soul and to the gods:

Like everything else, the soul consists of atoms, differing from other atoms only in being smaller and subtler; these are dispersed at death and the soul ceases to perceive ... The gods too are built of atoms, but they live in a less turbulent region, immune to dissolution. They live happy lives, untroubled by concern for human beings. For that reason belief in providence is superstition, and religious rituals a waste of time. Since we are free agents, thanks to the atomic swerve, we are masters of our own fate: the gods neither impose necessity nor interfere with our choices.

(p. 95)

It is clear how Epicurus can be seen as pious in relation to the gods by Diogenes Laertius. Epicurus acknowledges the existence of the gods, and believes them to live happy lives and free from decay. However, the religious find his conception of the gods as *material entities*, and the idea of their being indifferent to mankind and completely uninvolved in the world of humans, utterly unacceptable. To have them so completely written out of the story makes the assertion of their existence, in the eyes of the religious, utterly hollow. For all they matter to humans, they may as well not exist. Given the doctrine of the dissolution of the soul at death, this is for all intents and purposes a materialist, atheistic doctrine. This is why his enemies were so venomous in their attacks.

Epicurus thought that religion was responsible for the fear of death that troubles so many people. In particular the threat of hell, or the wrath of god or gods displeased with man's efforts in life, makes men tremble. As Kenny (2004) puts it:

> The aim of Epicurus' philosophy is to make happiness possible by removing the fear of death, which is the greatest obstacle to tranquillity … It is religion that causes us to fear death, by holding out the prospect of suffering after death. But this is an illusion. The terrors held out by religion are fairy tales, which we must give up in favour of a scientific account of the world.
>
> (p. 94)

Epicurus is undoubtedly pointing to a genuine social phenomenon, the inculcation of fear in followers of religions. However, it is only fair to point out that fear of death has a purely naturalistic explanation. Animals have no conception of death, but naturally fear danger. Humans, who come to recognise the reality of mortality, link death with danger easily enough, danger being, after all, some kind of threat to life. Notwithstanding this, Epicurus' argument rests on the assurance that on death, the atoms of which human beings are comprised – including the 'soul', which is, of course, composed of atoms – will disperse and will be gone for good. No thoughts or feelings, and no one to be subject to any kind of threat will persist. Of course, there are those who hear this account of non-being truly terrifying, but Epicurus would consider that fear irrational.

As with the teachings of the Lokayata, Epicurus believed the senses could be reliable as sources of information, but, as recognised by Democritus, false judgements may arise from observations. Reason must play its part in the process of gaining knowledge from perception. His ontological and epistemological theories in general are very similar to those of Democritus. In fact, Epicurus denied that he was a follower of Democritus, who lived 100 years earlier, but he must have been aware of his teachings, and their perspectives are in all essentials the same. One very important addition to the original atomic theory is the idea of *the swerve*, mentioned in the first quotation from Kenny above, and which contradicts Democritian determinism. (Epicurus also added weight to the properties of atoms.) Kenny explains:

> Nothing comes into being from nothing: the basic units of the world are everlasting, unchanging, indivisible units or atoms. These, infinite in number, move about in the void, which is empty and infinite space: if there were no void, movement would be

impossible. This motion had no beginning, and initially all atoms moved downwards at constant and equal speed. From time to time, however, they swerve and collide, and it is from the collision of atoms that everything in heaven and earth has come into being. The swerve of the atoms allows for human freedom, even though their motions are blind and purposeless ... The properties of perceptible bodies are not illusions, but they are supervenient on the basic properties of atoms. There is an infinite number of worlds, some like and some unlike our own.

(pp. 94–5)

The idea of the swerve is the first attempt to give an account of free will in a materialist universe. There are echoes of this idea in some twentieth-century philosophical speculations about a link between the quantum world and free will.

Epicurus sought a philosophy that would make happiness possible, and believed that the pursuit of pleasure is the key to happiness. The Epicureans shared with the Lokayata the accusation they were sensualists, hedonists. But what is known of Epicurus' teaching from Diogenes Laertius is very far from the hedonism of which he and his followers were accused. Kenny observes

Pleasure, for Epicurus, is the beginning and the end of the happy life. This does not mean, however, that Epicurus was an epicure. His life and that of his followers was far from luxurious: a good piece of cheese, he said, was as good as a feast. Though a theoretical hedonist, in practice he attached importance to a distinction he made between different types of pleasure. There is one type of pleasure that is given by the satisfaction of our desires for food, drink, and sex, but it is an inferior kind of pleasure, because it is bound up with pain. The desire these pleasures satisfy is itself painful, and its satisfaction leads to a renewal of desire. The pleasures to be aimed at are quiet pleasures such as those of private friendship.

(p. 95)

There is in fact much more to Epicurean ethics. He thought virtue was absolutely necessary to genuine pleasure. It is true that in his own life he shunned engagement in politics, but he believed people should live in society honourably.

From his deathbed, Epicurus wrote in a letter to Idomeneus,

I write this to you on the blissful day that is the last of my life. Strangury and dysentery have set in, with the greatest possible intensity of pain. I counterbalance them by the joy I have in the memory of our past conversations.

(Quoted in Kenny, 2004, p. 95)

Democritus and Epicurus are people who are free from the terrors of superstition. The Laughing Philosopher and the man who bears pain with extraordinary forbearance are at the heart of materialism's family tree, as is the Roman Lucretius.

Lucretius

Lucretius was a poet who lived from about 99 to 55 BCE. His place in this story is absolutely central, and yet he is not by renown a philosopher, but is acknowledged as one of the greatest poets of the Western tradition. This reputation is due to one epic work, De Rerum Natura – 'On the Nature of Things' – that is an exposition of Epicurean philosophy (Lucretius, 1997). However, some philosophers have argued that the poem shows Lucretius to have been an original thinker in his own right. According to Santayana and Bergson, he does not simply relay Epicurus' teachings in exquisite poetic form, but has a much deeper and more profound conception of the world of his own. The problem with this idea is that so little of Epicurus' work has survived, and so it is unknown what further works of Epicurus were available for Lucretius to draw on.

Lucretius is, like Epicurus, honoured by the existence of outrageous attacks from his opponents, of whom the most famous is St. Jerome. The saint tells us that

> The poet Titus Lucretius is born. He was later driven mad by a love philtre and, having composed between bouts of insanity several books (which Cicero afterwards corrected), committed suicide at the age of 44.
> (Quoted in Greenblatt, 2012, p. 53)

The authors of the Introduction to Lucretius' poem find this account dubious and doubt there is any evidence to support it. There are, though, plenty of pointers to it being an attempt by a 'holy man' to discredit an opponent. However, for a contrary view, see Gain (1969).

That Lucretius was indeed an enemy of the 'holy' is undeniable. In the poem, his hostility to religion is evident. Throughout it he mocks supernatural explanations and endeavours to find naturalistic accounts of many naturally occurring phenomena. He rejects religious accounts both of the origins of the universe and of inherent mind or purpose in nature.

De Rerum Natura is a long poem. It is divided into six sections, or 'books'. The first two describe the atomic theory of Epicurus, in which the universe is conceived as consisting of atoms and the void. Book 3 identifies the soul as composed of atoms and describes its dissolution on the death of the body. Book 4 is more concerned with epistemological issues, but in later verses discusses both dreams and sex. Book 5 describes the origin of the world and the dawn of civilisation, and Book 6 discusses various natural phenomena, including lightning, volcanoes, earthquakes and magnets.

Presented as the principal extant work in the history of materialism, it may come as something of a surprise that it begins like this:

O mother of the Roman race, delight
Of men and gods, Venus most bountiful,
You who beneath the gliding signs of heaven
Fill with yourself the sea bedecked with ships
And earth crop-bearer, since by your power
Creatures of every kind are brought to birth
And rising up behold the light of sun;
From you, sweet goddess, you, and at your coming
The winds and clouds of heaven flee all away;
For you the earth well skilled puts forth sweet flowers;
For you the seas' horizons smile, and sky,
All peaceful now, shines clear with light outpoured.

(p. 3)

This is the opening of a *poem*, not a philosophical treatise. No one can read the poem and come away believing Lucretius has expressed in these lines any kind of religious conception of Venus as a goddess with whom he is in a relation of worship. It is reasonable to think of this passage as expressing a love of natural beauty, personified in the figure of Venus.

The reader is soon made aware of the author's attitude to religion, and of his devoted admiration of Epicurus. After line 60, Lucretius writes

When human life lay foul for all to see
Upon the earth, crushed by the burden of religion,
Religion which from heaven's firmament
Displayed its face, its ghastly countenance,
Lowering above mankind, the first who dared
Raise mortal eyes against it, first to take
His stand against it, was a man of Greece.
He was not cowed by fables of the gods
Or thunderbolts or heaven's threatening roar,
But they the more spurred on his ardent soul
Yearning to be the first to break apart
The bolts of nature's gates and throw them open.

(pp. 4–5)

Lucretius can be very amusing in his critique of theorists. Here are some lines from his attack on early ontologists who thought the universe to be constituted by fire:

Of these the champion, first to open the fray,
Is Heraclitus, famed for his dark sayings
Among the more empty-headed of the Greeks

Rather than those grave minds that seek the truth.
For fools admire and love those things they see
Hidden in verses turned all upside down,
And take for truth what sweetly strokes the ears
And comes with sound of phrases fine imbued.
    . . .
To say moreover that all things are fire,
And nothing in this world is real except fire,
As this man does, seems utter lunacy.
He uses the senses to fight against the senses,
And undermines what all belief depends on,
By which he knows himself this thing that he calls fire.
He believes that the senses truly perceive fire,
But not the rest of things that are no less clear,
Which seems both futile and insane.

(pp. 21–3)

A champion of the materialist tradition could argue with some justification that materialist writers try to write clearly and plainly. Here Lucretius is criticising not only what he sees as a bizarre theory, but also the obscure and elusive language in which it is expressed.

De Rerum Natura is not an easy poem, but it rewards effort, and there is importance in the poem's artistic merit for the history of materialism. When the poem resurfaced after 1200 years of obscurity its quality as literature was vital in spreading its influence amongst the European intellectual elite.

The decline of materialist thought

Materialist thought in the Western world declined and descended into oblivion for over 1000 years. The reasons for this are varied and complex, but two distinct social trends are readily identifiable, one a general trend and the other more particularly associated with the Epicurean tradition. Both are associated with the rise of Christianity.

With regard to the general trend, it has been stated above that in the ancient world philosophy, science and religion were not distinct. The process of religion separating out from philosophy coincides with the growth of the monotheistic belief systems. The Jewish tribes of the Levant are identified as the source of the first great monotheistic culture, and it was from this tradition that Christianity emerged. A new attitude entered the intellectual scene.

Imagine two thinkers in dispute. Each tries to persuade the other of the correctness of his or her views, and to demonstrate the errors in the thinking of the other. This is the stuff of philosophical disputation. In contrast the two may hold a further attitude – that the view of the other is not only erroneous, it is *bad, wrong, dangerous.* It is his or her duty to convince the other of their error, because to believe as they do is, in some sense or other, *unacceptable. Toleration of the opposing view is replaced by its opposite* – *intolerance.* The tragic story of Hypatia and the library of Alexandria demonstrates its consequences.

Alexandria, the capital of Egypt when it was part of the Greek Empire, sometime in the third century BCE became the home of a great library founded by the Ptolemaic Kings. This library housed the cultural heritage of the Greeks, Romans, Babylonians, Egyptians and Jews. The goal was to attract outstanding scholars, philosophers and scientists, to create a great community of learning. It was a spectacular success. Amongst the thinkers who worked there were some of the greatest intellects of the ancient world, and the advances made in human knowledge were extraordinary. Greenblatt (2012) states that

> Euclid developed his geometry in Alexandria; Archimedes produced a remarkably precise estimate of the value of pi and laid the foundations for calculus; Eratosthenes, positing that the earth was round, calculated its circumference to within 1 percent; Galen revolutionized medicine; Alexandrian astronomers postulated a heliocentric universe; geometers deduced that the length of the year was 365.25 days and proposed adding a "leap day" every fourth year …
>
> (Greenblatt, 2012, p. 87)

For present purposes the crucial feature of the enterprise is its inclusiveness. The knowledge and worldviews of the whole world were within its scope. Dimitios, who was chief librarian until 284 BCE, was given a large budget and charged with ensuring that the library contained as many of the books in the world as possible and it is reputed to have held over half-a-million scrolls. A project to translate the Hebrew Bible into Greek was successfully undertaken there by seventy scholars commissioned by Ptolemy Philadelphus.

The seeds of the destruction of the Alexandrian culture were in the conflict between the traditional cults of pagan worship and the newer monotheistic cults. Greenblatt (2012) observes 'Centuries of religious pluralism under paganism – three faiths living side by side in a spirit of mingled rivalry and absorptive tolerance – were coming to an end' (p. 89).

The monotheists were the disputants alluded to above that were intolerant of the contrary views held by others. In the fourth century CE, Christianity achieved the status of official religion of the Roman Empire and in 391 CE the Roman emperor Theodosius the Great issued edicts forbidding public sacrifices and other pagan rituals. In the same year part of the library was destroyed on the orders of the Archbishop Theophilus. Within a quarter-of-a-century an event occurred that has been identified as the 'end of ancient world'. Violence between the pagan, Christian and Jewish communities grew in Alexandria. Iconoclasm broke out on a huge scale. Theophilus and his nephew and successor Cyril were avid leaders of the assault on pagans and Jews. Cyril demanded the expulsion of the Jews from the city, a demand rejected by Alexandria's governor Orestes, who although he had previously converted to Christianity resisted the Church having total control of the city. The city's 'pagan' elite also were opposed to the expulsion of the Jews.

Hypatia was the most distinguished member of the pagan elite. She is an extraordinary figure in history. She had a prominence unheard of for a woman in classical society based on her brilliance as an astronomer, mathematician and philosopher. She became the object of the Christians' wrath, and was accused of being a witch. In 415 or 416 CE a mob of Christians killed her with broken pottery and burned her corpse. Cyril was subsequently made a saint. Ironically, Hypatia's student Synesius became a bishop and incorporated neo-Platonist ideas into the doctrine of the Trinity. The whole intellectual tradition crumbled. The great collection of the library was lost to decay, worms and wanton destruction by the Christians.

The particular social trend that led to the virtual disappearance of the materialist tradition concerned the particular hostility the Christians felt towards Epicurean-ism. Traditions as powerful, intellectually rich and influ-ential as Greek philosophy were not going to simply vanish in the face of the rise of the Christian teaching. Rather, the Christians who emerged from a culture in which that philosophical tradition was dominant, over generations, incorporated certain aspects of Greek thought into the Christian canon. Features of the work of both Plato and Aristotle found a place in Christian doctrine. Besides anything else, they both believed in the immortal-ity of the soul. Stoic philosophy also appealed in certain aspects. But Epicurus had *nothing* to offer the Christians. The denial of Creation, the indifference of deities towards humans, the injunction to not fear death, the derision of religious ritual, the pursuit of happiness – all this was anathema. Epicureans were also loathed by pagans and Jews. Julian the Apostate, who tried to counter the rising Christian influence in the mid-fourth century, wanted to exclude consideration of Epicureans, and the Jewish authorities termed anyone who departed from the rab-binic tradition *apikoros* – an Epicurean.

Into a culture of mutual mockery came a two-pronged assault by the Christians. First, the great figures of the trad-ition, Epicurus himself and Lucretius, were subject to vilifica-tion – Epicurus as an advocate of excess – recall Horace's talk of his sty – and Lucretius a deranged suicide. Second, the mis-representation of the pursuit of pleasure as the pursuit of vice became central to the attack on the Epicurean tradition. As sin-ners, it is vice for man to pursue pleasure. Rather, we should respond to our deep guilt with due self-denial.

By the end of the first century CE *De Rerum Natura* was hardly read and its author had already begun a descent into philosophical oblivion. Thereafter the poem was lost, but not forever. Its revival is a remarkable story in the history of materialism. *De Rerum Natura* was the great flame of the materialist tradition, and therefore the bane of the Christian Church, and it was saved by the Christian monasteries.

In *The Swerve*, Stephen Greenblatt (2012) describes how in the early days of the establishment of the monasteries, monks were under an injunction to read. With the decline of the Roman Empire, the European intellectual tradition and institutions collapsed, and the survival of any learning at all was thanks to the monasteries in which monks learned Greek and Latin and were compelled to read. Armies of scribes over the centuries copied and recopied ancient texts, and the reading available for the monks went far beyond the Bible and the writings of the Christian fathers, and the pagan authors adopted by the Church – Plato and Aristotle.

After the great Italian poet Petrarch discovered in the four-teenth century a number of masterpieces by Roman authors who had lived more than 1000 years earlier, many others were inspired to go in search of lost works from the ancient world. One of these, Poggio Bracciolini, discovered, in an obscure German monastery, in 1417, a copy of *De Rerum Natura*. It is this event that Greenblatt identifies, in the title of his book, as an instance of an Epicurean swerve, because of the subsequent enormous influence the poem had on generations of European philosophers and scientists. More than that, and usurping the vocabulary of the opponents of materialism, it might be said that the reappearance of the poem occurred at that moment in history when the time had come for its mes-sage to be heard and its teachings absorbed. The story of materialism in the 400 years following the rediscovery of Lucretius' poem is the subject of the next chapter.

The triumphs of materialism: the
mechanical philosophy, the scientific
revolution and the Enlightenment

Three

Introduction

This chapter is concerned with the 400 years from the
rediscovery of *De Rerum Natura* in 1417 to the French
Revolution of 1789, though there is not much said about
the first century of this period. The 'triumphs' referred
to in the title of this chapter are not wholly attributable
to materialism, but it had a profound influence on the
thinking even of those who disavowed it. Materialist
thought is an important strand in the tumultuous centuries in Europe following the catastrophic, plague-ridden
fourteenth century. The reintroduction of Lucretius' poem
into European intellectual society acted like a shot of
adrenalin into the bloodstream of art, philosophy and science. The Protestant Reformation, the Renaissance and
the subsequent release of science from the constraints
imposed by the metaphysical teachings of the Church's
Aristotelian philosophy was followed by extraordinary
leaps in the understanding of the world, the ultimate
expression of which was Newton's *Principia* (1687). Ironically, Newton himself was neither a mechanical philosopher nor a materialist.

The rediscovery of *De Rerum Natura*

In *The Swerve*, Greenblatt (2012) documents the profound influence Lucretius had on many important thinkers of the sixteenth century. The poem brought a new perspective into European thought, after 1000 years of being totally dominated by Christian teaching that was founded on the works of Aristotle and Plato as well as the Bible, and which identified these sources as the truth and not open to criticism.

Many of those who were entranced both by the beauty and the ideas of Lucretius' poem were devout Christians who retained their religious convictions. Their capacity to hold what at first appears to be a paradoxical, if not contradictory, intellectual stance can be understood in the context of the interplay of two further traditions in Western philosophical and Christian thought. The first is a tradition of *scepticism*, mentioned in Chapter 1, which essentially challenges the grounds on which human knowledge of any kind claims to find support. The Greek Pyrrho and the Roman Stoic Sextus Empiricus are the two most famous classical sceptics, and their arguments were rediscovered in the same period as the re-emergence of *De Rerum Natura*. These ideas interacted with a longstanding strand in Christian thought that questioned the place of reason in religious belief, propounding the alternative view that religious belief was founded solely on *faith*. This stance came to be known as *fideism*. The upshot was that religion became detached from natural philosophy for many of the leading philosophers and scientists of the time, and all manner of speculations about empirical and metaphysical issues could be tolerated, provided they were not understood to challenge received religious doctrine.

One of the most influential of such thinkers was the French essayist Michel de Montaigne. His humanitarian open-minded attitudes are expressed in beautiful writings that were also a principal vehicle for the dissemination of Epicurean thought. The *Stanford Encyclopedia of Philosophy* gives this assessment:

> If we trace back the birth of modern science, we find that Montaigne as a philosopher was ahead of his time. In 1543, Copernicus put the earth in motion, depriving man of his cosmological centrality. Yet he nevertheless changed little in the medieval conception of the world as a sphere. The Copernican world became an 'open' world only with Thomas Digges (1576) although his sky was still situated in space, inhabited by gods and angels. One has to wait for Giordano Bruno to find the first representative of the modern conception of an infinite universe (1584) ... Montaigne, on the contrary, is entirely free from the medieval conception of the spheres. He owes his cosmological freedom to his deep interest in ancient philosophers, to Lucretius in particular. In the longest chapter of the *Essays*, the 'Apologie de Raymond Sebond', Montaigne conjures up many opinions, regarding the nature of the cosmos, or the nature of the soul. He weighs the Epicureans' opinion that several worlds exist, against that of the unicity of the world put forth by both Aristotle and Aquinas. He comes out in favor of the former, without ranking his own evaluation as a truth.
>
> (Foglia, 2014)

Montaigne's teachings on life emphasise the Epicurean injunction to not fear death. The essays read as startlingly modern and express a sensibility of gentleness and humility at odds with the brutality and ruthlessness of his times, dominated as they were by wars of religion. It is reasonable to link that sensibility with his love of Lucretius, from whom, Greenblatt informs us, he quoted nearly 100 times in the Essays. He is surely right to detect a 'profound affinity' between the two writers. He describes Montaigne as sharing

> 'Lucretius' contempt for a morality enforced by nightmares of the afterlife; he clung to the importance of his own senses and the evidence of the material world; he intensely disliked ascetic self-punishment and violence against the flesh; he treasured inward freedom and content. In grappling with the fear of death, he was influenced by Stoicism as well as Lucretian materialism, but it is the latter that proves the dominant guide, leading him toward a celebration of bodily pleasure.
>
> (p. 244)

Montaigne's *Essais* was one of Shakespeare's favourite books, and just as the literary qualities of Lucretius' poem facilitated the dissemination of its Epicurean ideas, so the literary qualities of Montaigne's essays helped disseminate the Epicurean perspective on the living of life.

Lucretius aroused the interest of many of England's leading writers of the time, including Spenser, Donne and the philosopher important in formulating the scientific method, Francis Bacon. The playwright Ben Jonson had his own (self-)signed copy of *De Rerum Natura*. It is worth mentioning too that a slightly earlier thinker of great subsequent influence was, according to some scholars, deeply influenced by Lucretius, albeit with a vastly different outcome from the fruits of Lucretius' influence on Montaigne. The following is taken from the abstract of a paper published in the journal *History of Political Thought*.

> If he had [a classical mentor], it was surely Lucretius, the leading Roman critic of ancient political idealism – a figure so important to the Florentine that in the late 1490s he copied by hand in its entirety De rerum natura and that he drew on it throughout his subsequent career. Machiavelli's republicanism is best understood as an appropriation, critique and reworking of ancient Epicureanism.
>
> (Rahe, 2007, p. 30)

As evidence for the claim of the importance of Lucretius in late fifteenth-century Florence, Greenblatt tells us that the deeply puritanical Dominican Savonarola thought it worthwhile to attack and ridicule the ancient atomists in a series of sermons. Within a few decades of its re-emergence, *De Rerum Natura* was seen as a threat to the established intellectual order.

The mechanical philosophy and Newton

The mechanical philosophers did not share a single conception of the workings of nature. Amongst their number are the substance dualist Descartes, the devout Christian Boyle and the materialist Hobbes. What they did have in common was a commitment to largely abandoning the Aristotelian theory of the four elements, the Ptolemaic theory of geocentrism and the magical ideas of the alchemists. All thought much was to be gained by thinking of nature as analogous to a clockwork machine whose hidden parts acted by unseen contacts. Most were atomists but Descartes denied there could be empty space and thought it was entirely filled by corpuscles of matter. The key problems were to understand what happens when bodies collide, what is conserved in mechanical processes and whether there can be a vacuum. Explanations of natural phenomena in terms of essences, ends, forms and qualities was to be replaced by quantitative and often geometrical descriptions. Major advances in scientific experimentation and instruments were accompanied by the growing consensus that scientists needed to investigate the behaviour of matter in precisely defined circumstances not readily accessible in the natural world, and so needed to design highly elaborate experiments using specialised equipment for this purpose (for example, using the air pump to create a vacuum to find out if sound needs the air to be transmitted). All agreed with the Democritean doctrine that honey atoms need not be sweet nor gold atoms gold in colour, and that such properties could be appearances caused by the motions and interactions of their atoms.

Amongst the most important mechanists were Marin Mersenne and Pierre Gassendi, both of whom were heavily critical of supernatural practices popular at the time, and made important scientific discoveries. The latter managed simultaneously to identify himself as a follower of Epicurus, and serve as a catholic priest. He wrote a commentary on Book Ten of Diogenes Laertius. He seems to have stayed clear of the wrath of the authorities, perhaps because of believing aspects of man to be immaterial. He taught Molière and Cyrano de Bergerac, who popularised the new natural philosophy.

Robert Boyle was a seventeenth-century scientist regarded today as one of the founders of modern chemistry, and an early pioneer of the scientific method. Although he was not the first to formulate the hypothesis, he is perhaps most famous for Boyle's Law, which states the inversely proportional relationship between the pressure and volume of a gas in a closed system maintained at a constant temperature.

Boyle was also a devout Christian, a Protestant, and made an endowment in his will for a series of lectures that came to be known as the Boyle lectures. The lectures were inaugurated in 1692 and their purpose was to defend the Christian religion against 'notorious infidels, viz. Atheists, Deists, Pagans, Jews and Mahometans'. The will stipulated further that the lectures should not address 'controversies that are among Christians themselves' (see Bentley, 1838).

Before going on with the story of the lectures, there are two comments to be made. The first is that, at the conclusion of the century, then, an eminent scientist saw reason to endow a lecture series in the service of defending Christianity. It seems fair to conclude that Christianity was seen by Boyle to be under threat. Regardless of the reference to people of other religions, it is clear that materialist thought had entered the mainstream of intellectual discourse. As claimed above in the previous section, this was in no small measure due to Lucretius, and devout Christians saw the threat such thinking posed.

The second comment brings to light the great paradox of the age. Boyle's passion as a scientist, and his success, depended on the materialist epistemology that found expression in the scientific method. The influence of the official science of the medieval Catholic Church, namely Aristotelian physics, had been abandoned. And yet, no tension was felt in him in maintaining a strong belief in god. Boyle's attitude is an example, then, of the great division of the mind that permits science to be pursued as an investigation of the natural world, conceived as free from the presence of any supernatural force in one state of mind, while a relationship to god can be maintained in another state of mind.

It needs to be understood that Newton and Boyle would have denied they suffered from a 'great division of the mind'. For them, natural philosophy is continuous with natural theology. However, their 'intelligent creator' god is very different from the god of the Old Testament. If god has become the Prime Mover, permitting the study of natural philosophy – in effect, science – to be conducted in the manner of a materialist atheist, with no reference to the supernatural, the expression 'division of the mind' does seem appropriate.

It would seem this was facilitated by a Protestant Christianity rather more than a Catholic one. The Reformation seems, in challenging and undermining the authority of the catholic authorities, to have granted licence to a much freer approach to the investigation of nature than had previously been possible. Earlier in the century Galileo had pursued scientific research while maintaining his Catholicism. But the findings that contradicted some biblical accounts of the nature of the world, and his being prepared to entertain Copernican ideas of a heliocentric universe brought him into conflict with the Catholic authorities, resulting in the threat of torture for heresy and a sentence of imprisonment at the pleasure of the Inquisition, which was quickly commuted to house arrest – detention in his own villa.

Returning to the story of the Boyle lectures, the first lecturer was Richard Bentley (1838), a brilliant young theologian and eminent scholar. The title page of the 1693 edition of the lectures reads

> The Folly and Unreasonableness of Atheism demonstrated from the Advantage and Pleasure of a Religious Life, The

Faculties of Human Souls, The Structure of Animate Bodies, and the Origin and Frame of the World, etc.

The last three of the eight lectures constitute 'A Confutation of Atheism from the Origin and Frame of the World', and address Newtonian Mechanics, the greatest scientific achievement of the age. Bentley engaged in a correspondence with Newton and there are passages from the letters that bear directly on the materialist tradition.

Bentley says that he is going to establish some propositions, one of which relates to gravitation and states

> That such a mutual gravitation or spontaneous attraction can neither be inherent and essential to matter, nor ever supervene to it, unless impressed and infused into it by a divine power.
>
> (p. 157)

The following passage occurs in the third of Newton's letters to Bentley.

> It is inconceivable, that inanimate brute matter should, without the mediation of something else, which is not material, operate upon and affect other matter without mutual contact, as it must be, if gravitation, in the sense of Epicurus, be essential and inherent in it. And this is one reason why I desired you would not ascribe innate gravity to me. That gravity should be innate, inherent, and essential to matter, so that one body may act upon another at a distance through a *vacuum*, without the mediation of any thing else, by and through which their action and force may

be conveyed from one to another, is to me so great an absurdity, that I believe no man, who has in philosophical matters a competent faculty of thinking, can ever fall into it. Gravity must be caused by an agent acting constantly according to certain laws; but whether this agent be material or immaterial, I have left to the consideration of my readers.

(Bentley, 1838, pp. 211–12)

It is interesting to see the name of Epicurus used in this context. The champion of Greek atomism is present in the intellectual ferment of the time, and in the opinion of the pious Newton taken to be trying to defend an absurdity – if there is nothing but atoms and the void, then gravity would have to be innate to the atoms, and so, granting the reality of gravitation, action at a distance will simply have to be acknowledged and accepted. Newton believes there must be something more, but is not prepared to affirm that this extra must be something immaterial; he keeps an open mind on the issue. So Newton does not question his Christian beliefs, but is unprepared to endorse openly an argument for the existence of god on the basis of his theory of gravity. More than 200 years later another great scientist did in fact give an account of the something more. The phenomena of gravitation, and the apparent action of one body on another at a distance, were given an explanation by Einstein's General Theory of Relativity. The 'something else' was the gravitational field of space-time. The gravitational field is, of course, neither material nor divine. It would not have served Bentley's purpose, then; there was no quick route to the demonstration of the existence of god. But it is also true, as discussed further in Chapter 5, that modern physics brought problems to traditional philosophical materialism.

Three key philosophers – Hobbes, Spinoza, Hume

To reiterate, it would seem that virtually all the great thinkers of the sixteenth, seventeenth and eighteenth centuries were familiar, to a greater or lesser extent, with materialist and atheistic ideas, and many of them first met these ideas in Lucretius' poem. Its atomic theory was compelling to many, both in its content and in the beauty of the form of its presentation. It is also true that virtually all the great thinkers of the sixteenth and seventeenth centuries believed in god. At least virtually all of them declared themselves to be theists, it being imprudent to do otherwise throughout the period. But for those strongly influenced by the materialist tradition, authentically retaining a religious belief system involved some radical alterations at the psychological level. It has been described above how for many scientists the mind had to be divided into two parts, the scientific secular part and the pious religious part. The findings of the new science contradicted not only Aristotle but also the Bible – Genesis states that the earth was created before the creation of the sun, and in Psalms it is said that the earth cannot be moved. While it would be at least theoretically feasible to abandon the pagan Aristotle, holding beliefs counter to the teachings of the Bible was inevitably problematic. The authorities did not approve of any of the various attempts at resolution, but some provoked greater wrath than others.

Thomas Hobbes was vilified as an atheist, because the god he claimed to believe in was a material one. To the

Christian, this is a heresy as bad as atheism proper, and of course suggests a familiarity with the teachings of Epicurus and hence familiarity with Lucretius. A minor episode in his story is that he is known to have failed in his mathematical project to square the circle. Failure was inevitable, of course – it is now known that the task of constructing a square of the same area as a given circle in a finite number of steps is impossible. But, in a metaphorical sense, the great thinkers of the age promoting science and the investigation of nature unencumbered by supernatural forces in nature and the looming presence of Aristotelian authority within, while at the same time seeking to uphold a religious doctrine with undeniable ontological claims, were all in an ultimately hopeless endeavour to square the circle.

The approach adopted by the great philosopher Spinoza was radically different, but brought even greater opprobrium. In contrast with the substance dualism of much medieval thought, which recognised just *res extensa* and *res cogitans*, extending things and thinking things, Spinoza believed there was just one substance, 'God', with infinite attributes, of which *thought* and *extension* are two.

It is perhaps necessary to give some justification for including Spinoza in this historical survey, as he was evidently neither atheist nor materialist. But the doctrine of the identification of the one and only substance as god, and the identification of god with nature, which became known as pantheism, was seen, not unreasonably, by the authorities as a step on the road to atheism, if indeed it had not already, in reality, reached that destination. Historically, pantheism can be seen to follow one path to nature worship, but it also points the way to another path of scientific exploration of nature unencumbered by any supernatural doctrine. The authorities were not amused.

Spinoza was a member of the Jewish community in Amsterdam, whose origins were in the Iberian Peninsula, from which they had fled to avoid persecution at the hands of the Inquisition. Spinoza was expelled from the Congregation of Israel in 1656. He was convicted of holding evil opinions – of believing horrible heresies – and his expulsion from the community was accompanied by a malediction that included these assertions:

> Let him be cursed by the mouths of the Seven Angels who preside over the seven days of the week, and by the mouths of the angels who follow them and fight under their banners ...
>
> Let God never forgive him his sins. Let the wrath and indignation of the Lord surround him and smoke forever on his head ... Let God blot him out of his book ...
>
> And we warn you, that none may speak with him by word of mouth nor by writing, nor show any favour to him, nor be under one roof with him, nor come within four cubits of him, nor read any paper composed by him.
>
> (Spinoza, 1967, pp. xxiii–xxiv)

This gives the modern reader some idea of just how scared the religious authorities were of thought beyond the limits prescribed by the teachings of the sacred texts.

A further reason for including Spinoza here is that he was much admired by Einstein. When asked to write a short essay on 'the ethical significance of Spinoza's philosophy', Einstein replied:

> I do not have the professional knowledge to write a scholarly article about Spinoza. But what I think about

this man I can express in a few words. Spinoza was the first to apply with strict consistency the idea of an all-pervasive determinism to human thought, feeling, and action. In my opinion, his point of view has not gained general acceptance by all those striving for clarity and logical rigor only because it requires not only consistency of thought, but also unusual integrity, magnamity, and – modesty.

(Jammer, 1999, pp. 44–5)

Spinoza made a major contribution to the liberation of science, beyond his pantheistic metaphysics. The focus in this survey has been on the irresistible rise, in the sixteenth and seventeenth centuries, of modern ways of thinking over medieval ones. To the modern mind, a claim about what the world is like is only to be taken seriously if the evidence and reasoning behind it can be tested. The modern mind is reluctant to accept any claim just because an authority has stated it. In modern times, of course, much is learned by instruction from experts and teachers, but those experts are only trusted to the extent that they are seen to operate in a culture of critical thinking, experimentation and theory testing.

However, it is also important to remember that the sixteenth and seventeenth centuries were violent times, and the traditional authority under threat was willing and able to guard its authority ruthlessly. When Descartes heard about Galileo's experience, he decided not to publish his work *Le Monde*, knowing that it would receive censure from the Rome, and censure could lead to persecution. In the work, he was going to present his theory of the world, including the heretical heliocentric theory, with the equally heretical implication that the earth moves, as a story, a fable, not about the actual world. But he was not a person to risks attacks of the sort Galileo suffered, despite him having presented his theory as 'merely' a hypothesis.

The seventeenth century witnessed extreme social conflict, most notably the catastrophic Thirty Years' War, responsible for up to 8 million fatalities. Superstition was widespread throughout European society, notwithstanding the advances in rationality amongst the intellectual elite, and this too was the source of murderous violence. The persecution of 'witches', which began in the fifteenth century and eventually fizzled out only in the late seventeenth century, is reputed to have seen the execution by burning or hanging of over 80,000 people, the greater number of them being elderly women. It is sobering to think that at almost exactly the same time that Descartes was getting frightened by the prospect of bringing the Church's wrath on himself by publishing *Le Monde*, a person accused of being a witch – in this case, a middle-aged man – was being tortured and then burnt at the stake in Loudun, a small town in France, just fifty kilometres, as it happens, from Descartes' birthplace.

In this social climate, most of those with doubts and criticisms of established authorities were effectively silenced by fear. Yet Spinoza's (2007) *Theological-Political Treatise* is written in a fearless voice. It was published anonymously, to save Spinoza from the wrath of the secular as well as religious authorities, though his authorship was not a secret for long. Its publication in 1670 was a critical moment in this story. In this work, much of it taken up with an analysis of biblical texts, Spinoza attacks the irrationality of traditional beliefs, denies the possibility of miracles and mocks organised religion and the biblical prophets. It was considered extremely dangerous. From the perspective of the rise of materialism, however, there is something more important than these attacks on established religion *per se*, which are presented, after all, as the work of a profoundly religious theist. For the story of materialism and the advance of science and the scientific method, what is crucial about the Treatise is that it constitutes a two-fold assault in the epistemological battle concerning evidence and the grounds for knowledge. On the one hand Spinoza dares to cast a critical eye on the sacred texts and subject them to damning scrutiny; the doctrinal claims are shown to be devoid of rational evidential grounds. On the other hand, he argues for the need for social and religious authorities to safeguard the individual's right to freedom of thought and freedom of speech, that 'Every man may think what he likes, and say what he thinks'. In other words, Spinoza articulates explicitly the critique that is implicit in the rise of rise of the scientific worldview. (See Nadler, 2013.)

The eminent philosophical voice outside France and Germany of the eighteenth century is that of David Hume. Hume is the most important philosopher of the period in the history of materialism. For many he is amongst the most important philosophers ever. Hume is the great sceptic in philosophy, and his scepticism is based on a radical empiricism. Like Pyrrho, he believes that all knowledge derives from the senses, and the senses are, evidently, unreliable. It is, therefore, reasonable to withhold assent to any assertion of knowledge. However, while it would be questionable to identify him as a materialist, his scepticism was less thoroughgoing than that of Pyrrho and he tended to a materialist worldview over others.

For an extreme sceptic, to declare there is no god is as unwarranted an assurance as to declare there is a god. But not all sceptics are so extreme and Hume's position can be misunderstood. It seems evident that Hume believed, for example, that his acquaintances were people who existed independently of him and had minds of their own. He believed the table at which he worked was a real object in the external world. Hume's scepticism consists in his denial that he is able to provide certain evidence for these beliefs. Hume's naturalism consists in his assertion that no philosophical arguments will prevent him having those beliefs in practical deliberation and everyday life.

There are important issues about the nature of knowledge at stake here. In contemporary analytic philosophy there is a much-discussed definition of knowledge as justified true belief. The issue then becomes what counts as justification, and in particular whether it requires certainty. Chapter 1 claims that the worldview of a society has a central role in maintaining a sense of social coherence and stability. To fulfil that role, it needs to be accorded a special status. It just does not seem feasible to say 'we have this worldview and they have that worldview, and the one is as good as the other'. Our one has to be better – more accurate, or true, and the grounds on which it is based, its justifications, have to be held fixed. The established authorities hated the sceptics for questioning the certainty of standard justifications and inviting reasoned debate about them, as much as they hated the materialists for their denial.

Hume posed a dilemma for the authorities. He was, first of all, an exceptional person of undoubted intellectual abilities. He went to university at the age of eleven. Throughout his life he was both given and denied positions of authority. He was not appointed a professor of philosophy because of objections from the religious authorities, but he was later appointed Secretary to the British Embassy in Paris.

It would have been absurd to doubt his genius. But in addition, he did not fit the image of the infidel held in the minds of so many religious people. Far from being unpleasant, or immoral, he was the most charming of men. He maintained friendships with people of profoundly differing opinions, and was widely admired. That such an eminent and exceptional man could hold such outlandish views must surely have afforded those views additional credibility.

There are two arguments he presented that are of particular relevance to the history of materialism, but consideration of one will be set aside until Chapter 4. His arguments against the reality of miracles has become justly famous, or infamous, depending on one's point of view.

Hume's arguments that cast doubt on the occurrence of miracles are powerful, if not conclusive. He was not denying the possibility of miracles. What he was questioning was their probability, and his argument amounted to saying that 'the balance of probabilities would always weigh against reports of religious miracles, so that accepting any such report would have to be a matter of faith rather than of reason' (Gottlieb, 2016, pp. 217–18).

The core of the argument is this: a miracle is a violation of a law of nature. A law of nature has been established on the basis of compelling evidence. It is probable that the violation did not occur and we are obliged to look critically at the evidence for the violation. In an Introduction to Hume's *Enquiry Concerning Human Understanding*, Millican gives a striking illustration to make Hume's case. He asks us to imagine he is worried about a very rare disease that affects only one person in a million, and that there is a test for the disease that produces positive results with 99.9-percent accuracy. That is, out of every 1000 people who test positive for the disease, 999 actually have it and one does not. He then imagines taking the test and getting a positive result. Ought you to conclude from this evidence that he probably has the disease? No, notwithstanding the positive result it is still very much more likely that he is the one in 1000 whose test result is wrong, than the one in a million who actually has the disease (Hume, 2007, p. l).

Reports of miracles are like positive test results for very unusual diseases. They must be treated with extreme caution, because however remote the possibility of error or deceit may seem, it may still be less remote than the possibility that a miracle has occurred. We should therefore adopt it as a maxim, Hume wrote, that 'no testimony is sufficient to establish a miracle unless the testimony be of such a kind that its falsehood would be more miraculous than the fact which it endeavours to establish'.

It can be argued that many of the great thinkers of the age asserted their religious convictions in bad faith, because it was prudent, and sometimes necessary, to do so in order to avoid persecution and 'the instruments'. The ultimate triumph referred to in this chapter's title captures this problem in a slightly different way. The treatise *The System of Nature* by Baron d'Holbach, published in 1770, is the first systematic attempt at an atheist materialist metaphysics and epistemology, together, inevitably, with a coruscating attack on the Church. The ironic fact is that d'Holbach published it anonymously. Materialism is only able to speak its name anonymously. Other pamphlets and books liable to bring their authors into conflict with the authorities he ascribed to people who were already dead.

d'Holbach was a figure in what Jonathan Israel (2002) identified as the 'Radical Enlightenment'. Many of the more famous figures of the Enlightenment, such as Voltaire and Hume, held beliefs radically unenlightened by our modern standards, but d'Holbach was in many ways the first to state, from a philosophical point of view, not only the essential metaphysical and epistemological materialist stances, but also its accompanying free-thinking conception of social ethics. He advocated freedom of thought and religion, and the separation of Church and State. However, it is also important to note that he was wary of teaching these theories to the masses (Herrick, 1985, chapter 5).

The legacy of the fourteenth century included the ravages of the Black Death, the devastating plague that carried off as much as a third of the population of Europe and the work of Plutarch, the great Italian poet who translated Virgil and gave impetus to the Renaissance. The rediscovery of *De Rerum Natura* was at the dawn of the fifteenth century. By the French Revolution at the close of the eighteenth century science had freed itself from the Church, and the age of the persecution of free-thinking by the Church had taken a more psychological form. In Europe, the threat and use of torture had by and large stopped. Science and technology began a march of progress of ever-increasing rapidity, having become the research and development department of the soon-to-be global economic system of imperialist capitalism. Untethering the epistemological instinct from the constraints of received doctrine liberated a force of such power that it offered a world of extraordinary wellbeing, and at the same time threatened the human race with catastrophic destruction.

Four

Introduction

The nineteenth century is a paradox in the history of materialism. On the one hand, there were many fruits of the triumph
of materialism discussed in Chapter 3, namely the liberation of
investigation into the nature of the world from the shackles of
religious dogma and argument from authority. The onward
march of scientific knowledge was evident. Maxwell's electro-
magnetism is a comparable achievement to Newtonian mechanics, and in different ways inaugurated the revolutions of
twentieth-century physics. There were also great advances in the
theories of heat and fluids, and importantly the Chemical Revolution of the end of the previous century spurred systematic
advances in chemical knowledge. Technological progress accelerated at a historically unheard-of rate. With regard to the latter,
however, the results could look decidedly mixed from some
vantage points. The dark side of this progress was the plight of
the poor in the industrialised nations, and the plight of the
peoples of the countries colonised by the European powers. The
consequences of these social trends have proven to be longlast-
ing and dangerous. The 'unholy' union of philosophical and
hedonistic materialism, if that is how the ethos of industrial
society can be characterised, deserves critical scrutiny. The high
ideals of the Enlightenment got lost somewhere along the way,

and indeed, as a European movement, the Enlightenment became associated on the world stage with the unenlightened realities of colonialism.

On the other hand, in the 'official' centres of philosophical inquiry – the universities – materialism became a minor voice. The whole European scene in philosophy was dominated by idealist thinkers. Stemming from Kant's magisterial *Critique of Pure Reason*, the main figures of nineteenth-century European continental philosophy, such as Hegel, Kierkegaard and Schopenhauer, are all hostile to materialism. The same is true in Britain, though the names of the leading philosophers of the late nineteenth century and early twentieth century in this tradition – Bradley, McTaggart and Alexander – are less well-known now.

Central to the story of the materialist response to academic idealism is Ludwig Feuerbach. A member of a group of thinkers known as the Young Hegelians, he turned the teachings of the master on their head and denied the primacy of the idea. For present concerns though, his significance stems from his influence on history's most famous materialist, namely Karl Marx. In fact, Marx's contribution to philosophical materialism is not great. His contribution, for good or ill, was to link the doctrine to a social movement that sought to release mankind from the horrors of industrialised society in its manifestation as imperialist capitalism. His Eleven Theses on Feuerbach is a vital text in the history of materialism, linking the philosophical doctrine with a hitherto-unknown militancy.

Feuerbach and Marx

Early nineteenth-century German philosophy is not only difficult for the lay reader to grasp. Scholars express profoundly differing interpretations of the texts and evaluations of the beliefs of the philosophers. It would be an unnecessary diversion to enter these disputatious areas in the context of the present purpose, but it is necessary to locate Feuerbach in the history of thought. As mentioned above, academic philosophy in Germany at the opening of the nineteenth century was dominated by idealism. Kant was, and is, a towering figure, and his critical philosophy had an influence it is difficult to overstate. A very different but also profoundly influential step in the development of idealism was taken by Hegel. Both of their philosophies were very much intertwined with the Christian tradition. Feuerbach was in his early adult life a follower of Hegel. The *Stanford Encyclopedia of Philosophy* reports that 'In notes for lectures on the history of modern philosophy that he delivered in 1835/36, Feuerbach wrote that idealism is the "one true philosophy", and that "what is not spirit is *nothing*"' (Gooch, 2016). However, he broke completely with Hegelianism and by 1844 Marx was able to state that 'Feuerbach is the only one who has a serious, critical attitude to the Hegelian dialectic and who has made genuine discoveries in this field. He is in fact the true conqueror of the old philosophy' (Marx, 1844, p. 64.)

The exact development of Feuerbach's thinking and beliefs is complicated and unclear. His writings seem to contain ideas at odds with one another. From the point of view of the history of materialism, what is more important is what Marx made of his writings rather than what ideas Feuerbach intended those writings to convey.

Feuerbach's contribution to the materialist tradition may be considered in two aspects: first a critique of Hegelian idealism, and second a critique of Christianity. He turns Hegelianism on its head and argues that rather than thought, or spirit, being primary, it is matter that is primary, and from which thought emerges secondarily. Correspondingly, he sees the Christian conception of god as a projection of human faculties. The *Stanford Encyclopedia* states

> [Feuerbach's] *The Essence of Christianity* is divided into two parts. In the first part Feuerbach considers religion 'in its agreement with the human essence' … arguing that, when purportedly theological claims are understood in their proper sense, they are recognized as expressing anthropological, rather than theological, truths. That is, the predicates that religious believers apply to God are predicates that properly apply to the human species-essence of which God is an imaginary representation. In the second part Feuerbach considers religion 'in its contradiction with the human essence' … arguing that, when theological claims are understood in the sense in which they are ordinarily taken (i.e., as referring to a non-human divine person), they are self-contradictory. In early 1842 Feuerbach still preferred that his views be presented to the public under the label 'anthropotheism' rather than 'atheism', emphasizing that his overriding purpose in negating 'the false or theological essence of religion' had been to affirm its 'true or anthropological essence', i.e., the divinity of man.
>
> (Gooch, 2016)

However accurate Marx's reading of Feuerbach was, he was certainly wrong about one thing – the 'old philosophy' was not overthrown, at least not in academic philosophy. The idealist Schopenhauer followed Feuerbach and became hugely influential. But for Marx that was not the important point. Having provided to Marx's satisfaction a theoretical rebuff to idealism, he then wanted to criticise Feuerbach, and all philosophy, for its essence as a reflective, unpractical activity – for, as he saw it, its fundamental passivity. He famously wrote his Eleven Theses on Feuerbach, the most famous being the last.

> The philosophers have only interpreted the world in various ways; the point is to change it.
>
> (Marx, 2000, pp. 171–4)

Marx was a philosopher who was also an economist and political theorist. He saw himself as in a radical tradition that included the social movement of working people, and he saw himself as in the tradition of radical hostility to religion. But his fundamental belief that the overthrow of an oppressive social system would only be possible by revolution introduced into the materialist tradition a new militancy. Recall the fear the materialists had of authority, and the way they felt obliged to disguise their true views, and, in the case of d'Holbach, to publish anonymously. With Marx all such timidity had gone and his radical political programme was publicly declared in *The Communist Manifesto* of 1848. His most famous criticism of religion, founded on his materialism and belief in the reactionary role of religion in society, was written a few years earlier in an introduction to a critique of Hegel.

> Religion is the sigh of the oppressed creature, the heart of a heartless world, and the soul of soulless conditions. It is the opium of the people. The abolition of religion

as the illusory happiness of the people is the demand for their real happiness.

(Marx, 2000, pp. 71–82)

There are major disagreements amongst scholars when considering the relation between Marx's beliefs and writings, and the subsequent twentieth-century movements that identified themselves as Marxist and which led to state power in some major countries. It is well known these states did little to advance the cause of real human happiness. The only point to be made here is that at some point in the communist tradition of materialism the view emerged that religion could be supressed by force. It was as though the materialists thought they could and would borrow the techniques of oppression that the enemies of materialism had employed throughout history. Besides the tragic break with the Enlightenment tradition of free speech implicit in this step, the crucial point is the lesson that was not learned was that this oppressive policy is bound to fail. Just as oppression could not kill off materialism, so oppression cannot stop religion. Indeed, all the great religions have a long history of oppression by other religions, and have established a noble tradition of martyrs. Besides being ethically reprehensible, the attempted suppression of ideas is psychologically stupid and doomed inevitably to failure.

Darwin

Of at least equal significance for the history of materialism is the impact of an advance in biology. It is difficult to overestimate the impact that the theory of evolution has had on our conception of the world, and difficult too to overestimate the scope of its acceptance in the scientific community. Dawkins' comment that 'it is absolutely safe to say that if you meet somebody who claims not to believe in evolution, that person is ignorant, stupid or insane', is perhaps more well known than these remarks in a BBC interview:

> ... I read Darwinism, and understood Darwinism at 16. And that was a big leap for me, because by the time I reached the age of 16, I had lost all religious faith, with the exception of possibly a sort of lingering feeling about the argument from design. So I'd already sort of worked out that there are lots of different religions, and they contradict each other, so they can't all be right – and that kind of thing. But I was left with a sort of feeling 'Oh well there must be SOME sort of designer, some sort of spirit which designed the universe and designed life.' And it was when I understood Darwin that I saw how totally wrong that point of view was, that rather suddenly scales fell from my eyes and I then became rather strongly anti-religious at that point.
>
> (www.bbc.co.uk/religion/religions/
> atheism/people/dawkins.shtml)

With a dramatic repudiation of the argument from design, and with a coherent theory about the natural evolution of the human species from the line of great apes, it felt as though key challenges to the materialist world view were cast into considerable, potentially decisive doubt.

This discussion recalls two great thinkers from Chapter 3. Newton, who had refused to point to the existence of gravity as grounds for a belief in god, nevertheless found the argument from design convincing.

Atheism is so senseless & odious to mankind that it never had many professors. Can it be by accident that all birds beasts & men have their right side & left side alike shaped (except in their bowells) & just two eyes & no more on either side the face & just two ears on either side the head & a nose with two holes & no more between the eyes & one mouth under the nose & either two fore leggs or two wings or two arms on the sholders & two leggs on the hipps one on either side & no more? Whence arises this uniformity in all their outward shapes but from the counsel & contrivance of an Author? Whence is it that the eyes of all sorts of living creatures are transparent to the very bottom & the only transparent members in the body, having on the outside an hard transparent skin, & within transparent juyces with a crystalline Lens in the middle & a pupil before the Lens all of them so truly shaped & fitted for vision, that no Artist can mend them? Did blind chance know that there was light & what was its refraction & fit the eys of all creatures after the most curious manner to make use of it? These & such like considerations always have & ever will prevail with man kind to beleive that there is a being who made all things & has all things in his power & who is therfore to be feared.

(Newton, 1710–)

It is worth observing how Newton sneaks into the theistic stance the need for the pious to be fearful. Religion has the aim of keeping men fearful. Epicurus, Lucretius and Montaigne seek the exact opposite. The materialist response to the above presented here is first Hume's response, and then the post-Darwinian response.

Gottlieb describes Hume as placing 'several layers of insulation between himself and his critique of this argument'. It was, frankly, sensible, not to be seen openly criticising the leading scientific thinkers of his time, nor to be seen casting doubt on what was taken as primary evidence for the existence of god. Hume is prudent and invents a 'friend who loves sceptical paradoxes', and has this friend make a speech 'on behalf of Epicurus'. The key point delivered in this camouflaged way is that there is an illegitimate leap from the evidence of the existence of a craftsman, an 'Author', to the assertion that that is also evidence for the existence of a god. Hume writes:

> If the cause be known only by the effect, we never ought to ascribe to it any qualities beyond what are precisely requisite to produce the effect: Nor can we, by any rules of just reasoning, return back from the cause, and infer other effects from it, beyond those by which alone it is known to us. No one, merely from the sight of one of Zeuxis's pictures, could know, that he was also a statuary or architect, and was an artist no less skilful in stone and marble than in colours. The talents and taste, displayed in the particular work before us; these we may safely conclude the workman to be possessed of. The cause must be proportional to the effect; and if we exactly and precisely proportion it, we shall never find in it any qualities, that point farther, or afford an inference concerning any other design or performance. Such qualities must be somewhat beyond what is merely requisite for producing the effect, which we examine.

Allowing, therefore, the gods to be authors of the existence or order of the universe, it follows that they possess that precise degree of power, intelligence, and benevolence which appears in their workmanship; but nothing farther can ever be proved, except we call in the assistance of exaggeration and flattery to supply the defects of argument and reasoning.

(Hume, 2007, Section XI, paras. 105–106)

As Gottlieb observes, from the nature of the eye and apparent signs of design, we cannot reasonably get to 'a being who has all things in his power & who is therefore to be feared'; he writes:

For all we can tell from his handiwork, the Craftsman may have had only limited powers over his materials. Perhaps he no longer has those powers, or has ceased to exist altogether, in which case there is little to fear from him. And any evidence of design certainly does not itself license the inference that the designer is or was supremely just or good.

(p. 216)

This is a powerful reply to the argument from design for the existence of god, but it wasn't until the theory of evolution that it was dealt a mortal blow. For with the advent of an understanding of the evolution of life it became clear that the evidence for a craftsman designer is indeed illusory. Dawkins communicates how evolution has created for those who observe the natural world the appearance of the work of a 'blind watchmaker', but given the essential conceptual ingredients of evolutionary theory – genetic variation and fitness to survive in the given environment – together with an unimaginably long time for the evolutionary processes to occur, there need be no agent acting as designer or craftsman at all.

This concludes the survey of the history of materialism. The analysis now turns to the philosophical details of a currently viable materialism, and the story begins with materialism's dearest friend, science, tolling the death knell of materialism as conceived in a continuous tradition beginning in India and Greece over 2000 years ago.

The evolution of materialism into physicalism

Part II

Five

Introduction

Materialism began as an imaginative vision about the nature
of reality and the true natures and properties of things.
While others speculated about the existence of gods, souls
and spirits, and imagined the material world was ultimately
composed of water, or fire or some combination of elements, the ancient and early modern materialist thinkers
envisaged a world in which the phenomena we observe,
and life itself, are produced by matter in motion in three-
dimensional space, and nothing more. This vision, in the
form of the mechanical philosophy, formed the intellectual
foundation for the transformation of natural philosophy
into natural science. The matter in question was thought of
as tiny particles, like motes of dust that are too small to
see. However, the idea that the motions of matter could be
explained in terms of action by contact, that is, one piece
of matter moving another by directly pushing it, did not
suffice. As discussed in Chapter 3, Newton's *Principia* of
1687, the founding theory of modern mathematical physics, posited a universal gravitational force acting between all
bodies instantaneously at a distance (Newton, 2016). This
seemed to many of his critics as mysterious as witchcraft,
but, as Newton said in reply, that is not so because it has a

precise mathematical form from which can be derived exact predictions. In due course his most diehard mechanist opponents succumbed to the empirical success of the laws of classical mechanics he formulated.

Materialism has always faced the challenge of explaining how matter in motion could give rise to ideas and sensations. However, it was not materialism's failure to account for life or mind that caused its radical revision. Ironically, it was the success of the science it did so much to inspire that undermined materialism. Materialism was made obsolete by developments in the area that materialists most admire, and for which they believed they were providing philosophical foundations, namely physics. In the centuries following Newton the study of matter in physics transformed our concept of matter. Our best current accounts are incomplete but are far removed from our concept of everyday material objects. This chapter explains briefly why the materialism of the ancients and early modern philosophers is no longer plausible.

Physics after Newton: the eighteenth and nineteenth centuries

Newton's great achievement was to unify the physics of the heavens and the physics of the world we see around us, but he certainly did not bequeath his successors a Theory of Everything. The idea of particles of matter subject to contact with other particles, and, subject to the laws of gravitation, was only going to get physics so far. Classical mechanics could be adapted to deal with fluids and continuous solids as well as rigid bodies like planets, but it was of next to no use in dealing with electricity, heat, light, magnetism and chemistry. Ontological issues were at the heart of science's endeavours to give an account of these phenomena. Exotic forms of matter, or semi-material stuff, or specifically non-material stuff were posited throughout the seventeenth and eighteenth centuries to provide the required explanatory accounts. To mention two: *phlogiston* was postulated in the 1660s as a bizarre fire-like substance that is present in combustible bodies and that could help explain combustion. *Caloric* was postulated in the effort to explain heat and was envisaged as a fluid that flows from hotter bodies to colder ones. Both these theories were eventually abandoned, the former in the eighteenth century, when Lavoisier found his classification of the chemical elements, and the latter in the nineteenth century, and physics eventually came to Francis Bacon's conclusion that heat is not a stuff but associated with the motions of matter.

The story of the historical development of the scientific account of the nature of light is much more complicated, and one in which materialism has an uncertain place. In the simplest terms, there was a contrast between *wave* theories and *particle* theories. Newton proposed a particle theory of light, but light has many properties that are better described in terms of waves, and if there is a wave, there is also medium that is vibrating. In the early nineteenth century the *optical ether* was introduced by Fresnel to support transverse waves of light. Meanwhile Faraday was unlocking the connection between electricity and magnetism and was the first to use the term *field*. Traditional materialism may not have much to say about waves, and is certainly challenged by the concept of the field that is somehow a medium but not quite material. When Maxwell introduced his electromagnetic field theory, the solid ether became something less substantial that obeys some very complicated equations and defies understanding in terms of ordinary matter. It permeates all of space and supports the wavelike propagation of light, and other waves, at a finite velocity. Magnetism is not understood as acting instantaneously at a distance, but at the same finite velocity as light and propagated by way of the field.

So traditional materialism can be seen to have mixed fortunes in the progress of science in the two centuries following Newton. Materialists had no single account of what kinds of substance make up the natural world. Nonetheless, the hope remained that all of physics and ultimately all of science could somehow be reduced to the behaviour of matter. Materialism had two great further successes. First, the kinetic theory of gases, according to which pressure, heat and temperature were the manifestations of atomic motions and collisions, led to substantial new mathematical physics and became the basis of our current understanding. Second, the triumph of atomism at the turn of the twentieth century turned it from a metaphysical theory into an experimental and practical science that unified chemistry and physics. Towards the end of the nineteenth century, many physicists believed they had come to the end of the road of discovery. The leading scientist Lord Kelvin is purported to have declared, around 1900, that 'There is nothing new to be discovered in physics now. All that remains is more and more precise measurement'. At this point, around 200 years after Newton's *Principia*, materialism, as an ontological theory, could stand quite confident in its foundational role for physics. It had been obliged to incorporate some new concepts, such as *gravitation* and *field*, but it did not feel as though these radically undermined materialism.

There were still 'atoms', and there was no recognition of non-material substances in scientific theory.

It is also pertinent at this point to mention the success of materialism in another crucial science – biology. There is a tradition reaching back to ancient times that tries to account for the apparently profound difference between living and non-living things. *Vitalism* is the name given to the tradition that postulates something non-physical to be found in, and only in, living things. It had its adherents well into the twentieth century but has fallen out of mainstream biology completely now. This will be referred to again in the discussion of the formulation of one kind of contemporary physicalism in Chapter 6.

The statement attributed to Lord Kelvin was, of course, profoundly mistaken. Very soon the discovery of radioactivity and the rise of quantum theory soon led to the radical transformation of our understanding of the nature of the atoms themselves. Far from being the indivisible building blocks of antiquity, they turned out to be compound entities whose behaviour and properties were beyond visualisation. Matter in the sense of extended stuff that takes up space like the familiar solid objects we see around us is, according to physics, not ultimately solid at all but mostly empty space. Matter is composed of atoms that are in turn composed of a nucleus and orbiting electrons. If an atom were the size of a football pitch the nucleus would be the size of the centre spot and the orbiting electrons on the touchline much smaller than that. But much worse than that was to follow in the twentieth century. With the discovery of subatomic particles, materialism still could feel there was a way of hanging on – still little things at the heart of everything. In the next section it will be shown that contemporary conceptions of the very small are incompatible with traditional materialism in any form.

In summary, then – having been the major intellectual stimulus to the scientific revolution, materialism gained its major triumph and readily abandoned its own speculative theories about lightning, volcanoes and other natural phenomena, safe in the belief that it had become the foundational philosophy of science, in both its ontological and epistemological aspects. But it didn't quite work out like that.

While materialist epistemology is the scientific method, developed and honed over the centuries, the application of that method of investigation has discovered a world of the very small of extraordinary complexity and strangeness, and thereby has condemned ontological materialism as a positive theory of what exists to the catalogue of false theories.

The twentieth century: physics transcends materialism

At its simplest, classical materialism imagined reality as a three-dimensional space, perhaps finite, perhaps infinite. Movement in time was an aspect of this reality, more or less imperfectly understood. Within the space were indivisible atoms, of varying kinds, and the void. All observable phenomena were ultimately composed of atoms.

Physics does not have a generally accepted theory of reality to contradict this, but there are general trends in twentieth-century physics that strongly suggest classical materialism doesn't really stand a chance. Rovelli (2014) offers an outline of current perspectives in physics that demonstrate this claim.

As mentioned above, advances in the nineteenth century introduced into physics the crucial concept of the field. There was, then, a need for something more than atoms and the void in order to provide a satisfactory description of reality. The electromagnetic field occupied space and could be described mathematically. Einstein took the concept of a field a vital step further. Recalling the struggle with the nature of gravitation discussed in Chapter 3 above, Einstein solved this problem not by identifying a gravitational field existing in space in the same manner as the electromagnetic field. He identified the gravitational field *as space itself*. Space is not a void with things existing in it. Space is not a void at all. Space is an entity that 'undulates, flexes, curves, twists' (Rovelli, 2014, p. 6). Space curves where there is matter. From these insights

> ... the magical richness of the theory opens up into a phantasmagorical succession of predictions that resemble the delirious ravings of a madman, but which have all turned out to be true.
>
> (p. 7)

Rovelli here is referring to, amongst other things, the fact that light doesn't move in straight lines but, rather, deviates; that time 'curves' as well – 'if a man who has lived at sea level meets up with his twin who has lived in the mountains, he will find that his twin is slightly older than him'; that when a large star burns up all its hydrogen it collapses in on itself and 'plummets into an actual hole ... the famous "black holes"'; that space is expanding, and that there was a beginning of the universe in which everything, including space and time, were compressed into an unimaginably small volume – the Big Bang.

Rovelli turns from the very large to the very small. He also moves from the extraordinary world of Einstein's General Theory of Relativity to the seemingly incomprehensible world of quantum mechanics. Another famous physicist remarked that if you are not baffled by quantum mechanics then you haven't understood what it is saying. From this it follows that the notion of clarification around this area of physics is itself compromised. But what is clear is that there is no comfort for the traditional materialist. This is because in quantum mechanics the very core ontological notion of objective existence is withheld from elementary particles.

Rovelli introduces a patchwork of ideas; Einstein showed light is made up of particles of light – photons. Einstein himself then struggles with the new path taken by the new theory. Neils Bohr

> understood that the energy of electrons in atoms can only take certain values, like the energy of light, and crucially that electrons can only 'jump' between one atomic orbit and another with fixed energies, emitting or absorbing a photon when they jump.

> (p. 14)

Equations were established in 1925 from which a conclusive explanation of the structure of the periodic table of elements could be derived – 'the whole of chemistry emerges from a single equation' (p.15). And then Rovelli introduces Werner Heisenberg:

> Heisenberg imagined that electrons do not *always* exist. They only exist when someone or something watches them, or better, when they are interacting with someone else. They materialize in a place, with a calculable probability, when colliding with something else. The 'quantum leaps' from one orbit to another are the only means they

have of being 'real': an electron is a set of jumps from one interaction to another. When nothing disturbs it, it is not in any precise place. It is not in a 'place' at all.

(p. 15)

Rovelli's chapter on *particles* may at first sight seem to bring some hope for traditional materialism. But in fact it is more like a further nail in the coffin of the cherished philosophical theory, *qua* ontological theory of what exists.

Light is made of photons; things are made of atoms; an atom consists of a nucleus surrounded by electrons; the nucleus consists of protons and neutrons; these latter are made up of quarks; the force that 'glues' quarks inside protons and neutrons is generated by particles that physicists, with a pleasing sense of the ridiculous, call 'gluons':

> Electrons, quarks, photons and gluons are the components of everything that sways in the space around us. They are the 'elementary particles' studied in particle physics. To these particles a few others are added, such as the neutrinos which swarm throughout the universe but which have little interaction with us, and the 'Higgs bosons' recently detected in Geneva in CERN's Large Hadron Collider. But there are not many of these, fewer than ten types in fact. A handful of elementary ingredients that act like bricks in a gigantic Lego set, and with which the entire material reality surrounding us is constructed.

(pp. 29–30)

From the standpoint of traditional materialism, so far, so good – seemingly a more sophisticated, more filled-out version of the original story. But these are not particles as intuitively understood, not very much at all like Lego bricks:

> The nature of these particles, and the way they move, is described by quantum mechanics. These particles do not have a pebble-like reality but are rather the 'quanta' of corresponding fields, just as photons are the 'quanta' of the electromagnetic field. They are elementary excitations of a moving substratum similar to the field of Faraday and Maxwell. Minuscule moving wavelets. They disappear and reappear according to the strange laws of quantum mechanics, where everything that exists is never stable, and is nothing but a jump from one interaction to another.
>
> (p. 30)

Evidently, the particles of contemporary physics are of a qualitatively different kind from the particles – the atoms – of traditional philosophical materialism.

Atoms in the void? Forget it. Materialism, as encountered so far in this story, has been left far behind not only by the extraordinary complexity of modern physics, but also by its theoretical and mathematical sophistication. In the space of 300 years, there has been an exponential growth in scientific and technological understanding, in the course of which reality has revealed itself to be more bizarre than could possibly have been imagined.

The challenges of contemporary physics

Physics now rests on two pillars, Quantum Theory and the General Theory of Relativity. These are extraordinary achievements of the scientific enterprise, with vast explanatory power and with startling degrees of confirmatory verification. The problem is that it is not known how to combine them, or even if they can be combined. A large proportion of theoretical physics in the early years of the twenty-first century was devoted to the project of finding a consistent unified theory that incorporated the essence of both great theories.

In general terms, classical materialism was, inevitably, couched in the language and concepts of everyday life. The advances in physics in the last 150 years have exposed the ultimate inadequacy of everyday concepts to convey the discovered world of the very small and the very large. The complexities and intricacies of processes at the large and small scale cannot be accurately described with concepts of ordinary language. And this perhaps is not surprising, given that it is a language that has evolved, after all, in the context of human beings striving to live in the world of medium-sized physical objects. Our scientific image of the world is one from which mathematical representation is inelimin-able. This applies to all the sciences, not just to physics.

If there is one clear lesson from the history of physics, it might be this – take nothing for granted. At this point in time physics lacks a unified theory of reality, but suppose it found one – would it then be in a position to answer the ontological question?

Physics would have *an* answer, but it could only claim to be *the* answer if the theory were shown to be, in some important sense, the *final* theory. This is because the history of science is a story of one conception of the world holding sway until another theory, superior in some way or other, takes its place. Unless a point is reached in the future when science can argue convincingly that it has found the *Theory* of *Everything*, any answer science gives to the ontological question will be provisional – 'in our present conception of reality, or according to our best theories so far, so and so exists'.

This chapter recounts a sobering story for traditional materialism. However, all is not lost. If it has been necessary to lay to rest the theory, *qua* ontological theory of what exists – as was stated above – the implied *negative* claim of the theory, about what *doesn't* exist, survives. What theoretical physics has not brought into the picture is anything answering to the description of the spiritual or the divine. The next chapter turns to the responses of philosophers in the materialist tradition to these developments in physics.

Six

Introduction

Philosophers in the materialist tradition had little option but
to accommodate the new directions that physics had taken in
the twentieth century. It became clear that ontology – discovering what exists – is a matter for science, rather than philosophy. Philosophers have played a part by bringing some
conceptual clarification to the bizarre and paradoxical world
of current physics. Such philosophical work, to be valuable,
has to be predicated on an understanding and appreciation of
the physics, and to that extent it is philosophy undertaken
within physics. Conceptual clarification is a necessary part of
physics.

Most philosophical work in the materialist tradition since
the rise of the new physics has focused on the nature of
psychological phenomena, these being the greatest challenge
for materialism. There is a vast and growing literature in
this area of metaphysics, but it is possible to identify two
central and very different approaches to the challenge. Both
have at times adopted the new name *physicalism* in place of
materialism, but with radically different ideas about how
this title is to be understood.

They can be initially identified as *reductionist* and *non-
reductionist* approaches to psychological phenomena. Reductionist

approaches attempt to demonstrate that, appearances notwith-standing, psychological phenomena are in fact material. Non-reductionist approaches deny the feasibility of the reductionist approach and seek to retain the spirit of materialism while acknowledging the existence of non-material psychological phenomena. In this chapter a representative example of each approach will be described.

The reductionist theory is known as *mind-brain identity theory*, or *mind-body identity theory*, and hereafter in this chapter will be referred to simply as *identity theory*. It wears its nature on its sleeve. It seeks to say, in some way or other, the mind and the brain are the same thing. This theory is examined in the next section.

The non-reductionist theory can be called *supervenient physicalism*. Like identity theory, it makes some quite radical claims and in many ways it can be seen as *less* like traditional materialism than identity theory, but *more* in tune with contemporary physics. Both theories require a confrontation with the bizarre, and both theories leave some cherished beliefs about human beings under a cloud of doubt.

From the perspective offered here, one approach is considered more successful than the other. Identity theory has, to all intents and purposes, had its day, if a strict accord with modern physics is taken as a requirement of physicalism. However, this is a statement that will be considered contentious by many. Furthermore, the central concept of the other approach, *supervenience*, has its own critics with regard to its theoretical value. It can be challenged as offering more mystery than explanatory power.

So, in summary, the path divided, but as physicalists, as inheritors of the materialist tradition, the central preoccupation for both traditions was the characterisation of the apparently non-material phenomena of the world. And the seemingly non-material phenomena of greatest concern were the psychological phenomena of our human experience. While the approaches differ profoundly, neither one provides much solace for our cherished self-image. The human being who emerges in these physicalist accounts may not feel to be quite what human beings imagined themselves to be previously.

Mind-brain identity theory

The discussion draws on the introductory essay in *Modern Materialism: Readings on Mind-Body Identity*, edited by John O'Connor (1969). The book contains foundational papers in identity theory, not only by the founders of the tradition, but also by some of the biggest names in twentieth-century analytic philosophy. This shows how central this project was for analytic philosophy in the latter part of the last century.

The introduction clearly describes the landscape in which this theory has been developed. It is presented in three principle sections. The first, 'Man and Nature', makes clear that in parallel with the more general materialist perspective, identity theory focuses specifically on the question of the nature of the human being. O'Connor poses the rather vague question 'Is man like or unlike the rest of things in the world?' as a starting point for contrasting the materialist identity theory with the non-materialist view discussed here in earlier chapters that sees the human being comprised of a material body and a non-material soul or mind. All materialists will concur with O'Connor in seeing man as 'like the rest of things in nature', but seriously question the following step in his argument:

... science is more and more able to explain so-called mental phenomena on the basis of physical laws, as is shown by recent work in the chemical basis of schizophrenia, for example. They also note that computers seem more and more capable of approximating what men call rationality. On this basis, [materialist] philosophers conclude that an adequate account of human beings can be given in terms of sciences that seem to be capable of producing adequate accounts of the rest of nature.

(p. 4)

Clearly much hangs on the notion of explanation, but nobody today would seriously suggest that finding neuro-physiological correlates of the symptoms associated with schizophrenia could be described as an *explanation* of schizophrenia. Equally, O'Connor employs the rather suspect notion of adequacy – what are the criteria for adequacy of an account of some phenomenon?

The second section, 'Physicalism', introduces O'Connor's understanding of how materialism found itself obliged to undergo a change of name. Describing materialism as the view

(1) That man is like nature in that both consist of the same materials – those materials studied by natural science, in particular physics – and (2) that an adequate account of human beings expressed in terms of a scientific theory is a complete and adequate account.

(p. 5)

In the past this was called materialism, but

> in its modern form, it has become known to philo-
> sophers as *physicalism*. This is perhaps a better name for
> it, since 'materialism' suggests that everything is made
> up of matter and contemporary physics has shown
> that there is a lot more to the world than matter;
> indeed, the line between matter and force is anything
> but sharp. The word 'physicalism', on the other hand,
> seems more flexible; it merely suggests that man,
> whatever he is, can be described adequately with the
> terms and concepts employed by the science of
> physics.
>
> (p. 5)

The suggestion seems to be that the change of name does
not signify anything more profound than a recognition that
physics recognises more in reality than matter. The lessons
from physics are limited to this. But referring back to Chapter 5, it will be recalled that the revolution in the understanding of the world arising from the discoveries of
twentieth-century physics has more far-reaching consequences than this.

O'Connor goes on – the 'program' of physicalism involves
two tasks: first to 'indicate what form a physicalist account of
human beings might take', and second 'to demonstrate that
there is no difficulty in principle in the adoption of physicalism'. With regard to this second task, O'Connor writes

> [the physicalists] are concerned to show that physical-
> ism offers at least a possible account of human
> beings ... the chief objection to physicalism is that it

offers no way of accounting for the mental life of human beings. Minds are not physical, it is argued, and consciousness is not a physical property. Since human beings have minds and are conscious, physicalism is bound to be an inadequate view. Given the seriousness of this charge, it is not surprising that philosophers who are sympathetic to physicalism spend much of their time and effort attempting to show that the charge is not true.

(pp. 7–8)

O'Connor cites three ways physicalists characterise mental phenomena – the first claims that all mental terms stand for entities, and these entities are physical; the second, known as eliminitavism, denies there are any such things as mental entities at all – 'to have a mental life according to this theory is just to be disposed to behave in certain ways in certain circumstances'. The third, generally preferred by the majority of identity-theory physicalists, finds a path between these two extremes. A feature of such a theory would be to understand a person using mental terms as 'speaking of experiences that human beings have but not of "mental entities" that human beings experience':

There are no such things as pains in the world, but there are experiences we call 'being-in-pain.' (The physicalist might then go on to say that the experience of being-in-pain is really a physical process.)

(p. 9)

In addition to an account of the mental, an account of the physical is required also, in the specific context of groundwork for the identity theory. O'Connor claims that most contemporary physicalists hold that

> The physical aspect of man that is relevant to an account of his mental life consists of both his body (considered in macroscopic behavioural terms) and his brain and central nervous system. These philosophers believe that if some mental terms can be analysed in terms of dispositions to behave in certain ways and others can be understood as referring to experiences that are really physical processes, then the physicalist position is in very fine shape.
>
> (p. 11)

O'Connor's third section, 'The Identity Theory', highlights how complicated the philosophy of identity can be. In particular, four issues arise. The first is to ask if the identity mooted is necessary, like '2 + 2 = 4' and 'All bachelors are unmarried', or contingent, like 'The table is made of oak'. Identity theory is bound to deal in the contingent kind of identities; its claim cannot be based on the meaning of terms.

Second, there is some dispute as to whether the identities of concern are arrived at by a process of discovery or a process of decision. It would seem that identity theory implicitly imagines an identity to be discovered, but it may too be hard to conceive of a 'discovery that could show that things such as experiences and brain processes are identical, as opposed to being merely constantly correlated' (p. 13).

Third, and much more contentious than the first two, concerns the strength of the identity claim. O'Connor describes *strict identity* only if two things share all their nonmodal properties. But something here won't do — strict identities involve only one thing, perhaps going by different names, or different descriptions. The famous example in philosophy is 'The Morning Star is the Evening Star', as the descriptive terms both refer to Venus. Science poses a difficult version of this issue — is the table identical with the cluster of atoms that comprise the table? The table is still but the cluster is in constant motion. The identity theorists may argue on this point:

> This has led some philosophers to say that the sort of identity that might suffice for a physicalist is some sort of theoretical identity, an identity based upon a scientific theory ... Whether an identity can be defined this loosely and yet actually remain an identity is a moot question. Some philosophers feel that anything short of strict identity would be merely a correlation of two things, and this would not be sufficient for an identity theorist.
>
> (p. 14)

The fourth, equally important, issue concerns the question of whether the identities are general or particular. In philosophy this distinction is often identified as a choice between *token* and *type* identity. So the issue is this — is identity theory making a claim that there is a token identity statement that can be made linking a particular mental entity, event or process with a physical entity, event or process (that is, they are the same thing under different descriptions), or is the stronger claim being made that mental entities, events or processes of a given type A are always identical with a physical entity, event or process of type B? It would look as though identity theorists would want to make the stronger claim, but it may be possible to develop a theory that includes both general and particular identity statements.

O'Connor next turns to the question of the precise detail of the identity claim – what is being said to be identical to what? By far the most widely accepted answer is that there are identities between experiences and brain processes:

> According to this theory, there are no mental entities, but there are experiences – for example, having an after-image. This experience is identical with some brain process. Which particular process it is may require a lot of neurophysiological investigation to discover, of course. The identity theorist does not claim merely that whenever a person has a certain after-image he is undergoing some brain process, for this view could be held by a dualist. Rather he claims that the experience is the brain process. There is one thing and not two.
>
> (p. 16)

This is, in essence, the perspective offered of identity theory, and its merits and deficiencies need to be judged in order to ascertain its place in the long history of materialism.

It does, at first sight, to be a rather strange theory. At first sight, it would seem to be obvious that mental experiences are not brain processes; they are, evidently, different types of thing. There may be correlations between mental experiences and brain processes, but to make an identity claim seems profoundly counter-intuitive.

Of course, identity theorists recognise that the theory involves a radical challenge to intuition, and while it seems these intuitive criticisms of identity theory have considerable force, identity theorists would argue vigorously that these objections can be met. The arguments are not pertinent to the present purpose, but there is another kind of criticism that would seem to be potentially decisive.

To go back to square one, the point of identity theory is to defend materialism, and the way is to show that apparently non-physical things are in fact physical after all. And the things the apparently non-physical things are identified with are brain processes. But what is the essence of the materialism that is being defended? Originally it was some version of an ontology of atoms in the void, but this has had to be abandoned, because materialism has always been obliged to take the lead from science, as scientific discovery arises from the same epistemological methodology that materialism espouses, and science has abandoned the atoms in the void ontology. What the identity theorists fail to take on board is just how profoundly science has abandoned traditional materialism. They talk as if brain processes – dynamic systems of neuronal networks – carry, for science, the full ontological status once ascribed to the atoms in the void. They imagine that the atoms and molecules that in composition form the brain processes supposedly identical with mental experiences are given by science unambiguous status as entities that exist. Indeed, science does acknowledge their status as things that exist, but no more than thoughts and feelings; the implicit ontology of contemporary physics recognises these mental things as existing as much as atoms and molecules and neurones.

This is because these phenomena are not the fundamental elements of reality. In fact it is not known what these fundamental elements are, or what they are like. Taken at one level of scientific scrutiny a world of atoms and molecules is discovered and described. At another – higher – level psychological phenomena are found. In the search for the deepest structures, work goes on – in quantum gravity, in string theory and perhaps other as-yet-unsung approaches.

A leading contemporary philosopher, Tim Crane, has written in the *Times Literary Supplement* some words the defenders of materialism are obliged to take seriously.

> We know, with as much certainty as we know anything, that we have conscious thoughts and experiences; we have memories and dreams, we imagine, desire and regret things; we plan our actions and form intentions; we form emotional attachments and structure our lives around them. All of these things are underpinned, in a way that we do not yet understand, by the unbelievable complexity of the brain and its mechanisms, some of which extend into the body. We need to make connections between the knowledge that we have about our minds and the knowledge that we have of our bodies and brains: but we do not yet know how to connect this knowledge in a systematic, illuminating way.
>
> In order to make these connections, we must first accept the irreducible reality of the mental, or psychological, for what it is. To connect two things, these things must both exist. Psychological reality is not a separate 'substance', and it is not just matter either. Our psychological states and processes are as real as anything going

on inside us – as real as our weight, our metabolism, our body temperature – and the fact that they are invisible is no more an objection to their existence than the fact that our weights and temperatures are invisible is an objection to theirs.

Some will protest that this is all very well as a way of speaking, but in reality 'all there is' to the mind is the brain, and this psychological talk is just that: talk. Sometimes you hear scientists saying that psychological reality is just another 'level of description' of the brain. But this is sloppy thinking: our dreams, experiences, thoughts and intentions are not 'descriptions'. They are events or processes going on in us, as real as the neural activity with which they are correlated.

(TLS, 26/5/17p.8; emphasis added)

It may be surprising that this clear, seemingly obvious account of the situation needs to be stated at all. But it does – in some philosophical and scientific circles versions of identity theory remain influential, despite their flaws, and despite being in conflict with the perspectives of contemporary science and contradicting the evidence of our experience as human beings. Crane sees himself arguing against materialism, but in the next section it will be shown that a coherent physicalism can be expressed that is true both to the spirit of materialism and to every word in the quotation from Crane.

Supervenient physicalism

There is a considerable amount of work in contemporary analytic philosophy on non-reductive formulations of physicalism. The authors of the present work have offered one formulation, and it is this that is presented here in outline (Brown & Ladyman, 2009). For readers wishing to explore the field, Justin Tiehen's paper 'Recent Work on Physicalism' (Tiehen, 2018) provides a very useful summary of the competing formulations, together with numerous references.

The presentation of supervenient physicalism will be approached in stages. The first stage, elements of which will be modified in the final formulation, begins of necessity with supervenience and its associated concepts.

From here on, discussion of *matter* will be replaced by discussion of *the physical*. The physical is to be understood as *that which is referred to in theories of physics*. These theories do not, and it is forecast will not, ever posit psychological or spiritual phenomena.

The concept of supervenience is central to twenty-first-century physicalism, and while it is not a particularly complicated idea, it can nevertheless be difficult to grasp. It is a concept that is embedded in a broader theoretical perspective, and it is going to be helpful to outline certain fundamental ideas on the basis of which supervenience takes its role in physicalist theory. Because of the limitations of space, the following is inevitably a rather impressionistic account of the field, but it should be sufficient for an intuitive grasp of the physicalist world view.

(a) Change. The notion of change is widespread in everyday discourse, and for present purposes there is no need to elaborate the basic idea. It is a familiar idea to think that a thing can change in some way or other and still retain its identity as that thing. The table can become wobbly, but it is still that table. A person can grow old, or become more placid, while still being that same person who was there before the change.

There is one important distinction philosophers draw between kinds of change, and this relates to the distinction between internal and external properties. The table may have the internal property of being made of teak, and the external property of being five metres from the bookcase. If the bookcase is moved the external property mentioned will change – the table will now be six metres from the bookcase, although its internal properties will have undergone no change. The primary concern here is with internal properties.

(b) Levels of reality and the hierarchy of sciences. These issues are very complex and contentious. But one clear lesson learned from the scientific revolution of the past 500 years is that reality is much more complicated than it may be imagined to be from basic observation by way of the senses. The atomists' idea that all material things are composed of tiny material things has been completely superseded by the rise of atomic and molecular chemistry. As was described in the last chapter, the concepts used in the description of the world of everyday things – tables and trees, people and lions, buildings and cars, basically of medium-sized objects occupying space – are unsuited for the purpose of describing the atomic and subatomic world, as discovered through the advances of theoretical physics and observed through the senses augmented by scientific instruments.

Yet in some way that bit of reality identified as the table is also an object that can fall under the description offered by the physics of condensed matter in terms of atoms and molecules. Reality can be conceived of as *layered*, and there is a hierarchy of human theories appropriate to the corresponding layers of reality.

Of course, in some basic sense, reality is not layered, it just *is*. It is conceived of as layered from the viewpoint of human knowledge and understanding. Human knowledge brings an understanding of the table – its weight, its material constituents – teak, for example – its method of construction, its use and purposes. But human science also provides information with regard to its constituent atoms and molecules, its composition in terms of elements and the way that it exhibits solidity to human touch. And theoretical physics is preoccupied with bringing an understanding of the *sub-atomic world* underlying the atomic and molecular structure of the table. Furthermore the hierarchy itself is complex and simple models of the reduction of the sciences to physics have also become completely outdated.

(c) The special sciences. Physics is taken to be the most general science that concerns itself with all phenomena. In contrast, consider sciences such as economics, or biology, or psychology. These sciences concern themselves with a subset of real phenomena. Biology is the science of life; its concern is the nature of living things. Economics is concerned with economic phenomena – markets, money supply, interest rates, levels of supply and demand, gross national product. It enters the realm of psychology, the science of psychological phenomena, in its concern with the decision-making processes of consumers. These are the behavioural sciences.

The suggestion here is that there is a hierarchy of sciences. It is not a completely neat structure; the special sciences overlap and share concerns, but physics is seen as the foundational science upon which the others are built. And physics itself can be seen as layered. The mechanics of solid objects is part of physics. The term *fundamental physics* can be used to identify the work seeking to understand and describe the more fundamental structures of reality.

The consequences of the layered picture of reality are significant. Consider our everyday 'science' of the objects in our houses. According to this theory, tables are solid and will support the dinner plates. What is discovered at the atomic level is that the table is not solid; the atoms on which it is composed are not dense in any everyday understanding of density. The slice of space-time housing an atom is largely empty. However, atomic theory provides an explanation of why the table supports the dinner plates. An account is given of how a network, or web, of atomic forces creates the property of solidity – that is, the property of supporting the dinner plates. Atomic science and table science both acknowledge the property of supporting the plates, but they each have different stories to tell about the nature of that property. Table science says it is because the table is dense; atomic science says it is not dense but held together by powerful atomic forces.

On the basis of these ideas, the concept of supervenience can be described. Here is a provisional definition:

A supervenes on B if there can be no change in A unless there is a change in B.

A and B are left undefined; they may be properties or things, or even levels of reality. Referring again to the table, there can be no change in any feature of the table observed by the dinner guests without there having been some change in the subatomic structure of the table. If the table has become warmer, the atoms *couldn't possibly* be just as they were when the table was cooler. To expand on this idea, consider the following statements:

The beauty of a work of art supervenes on its material form.

It is not possible to imagine a situation where two pictures are identical except that one is beautiful and the other is not. Notice that a strong concept is involved – 'identical' means what it says. Now, it may be argued, two things can't be truly identical. Referring back to the paragraph on change, if they are separate entities, they are going to have differing external properties, for a start. But they are also going to have different internal properties; if they are paintings, they will be on different canvasses; if sculptures, they will be made of different pieces of stone.

These objections are perfectly reasonable, and philosophers draw on further complex ideas to circumvent the objections in order to convey the idea. So a philosopher may imagine our reality, R, and a separate possible world, W, like R in all respects except that the painting is beautiful in R and not in W. And he may claim that W could not possibly exist. Notice this is not about whether people think the painting is beautiful or not; it is about whether the painting is beautiful or not. Now a step further:

The mental phenomena of an individual supervene on her material constitution.

It is not possible to imagine that two people are alike in every way – in particular, the material constitution of their bodies is precisely the same – and yet one thinks the painting is beautiful and the other does not. Or one is happy and the other is sad. Once again, the philosopher would turn to possible world analysis to render this idea coherent, but the point is made with regard to the concept of supervenience – there is a fundamental dependence the supervenient entity has on the underlying, *subvenient* entity.

It is at this point that the place of supervenience in physicalist theory may become clearer, for it is evident that if mental entities are supervenient on non-mental, physical, subvenient phenomena, the existence of a soul, or spirit, independent of, autonomous from, the body, is thereby denied.

If the lessons of contemporary physics are incorporated into this perspective, a more general statement can be made sufficient for the purpose of expressing physicalism.

Any non-fundamental property, event or entity supervenes on the subvenient reality underlying it.

As an instance, all the everyday, medium-sized objects that are familiar to us from our ordinary perception of the world supervene on the underlying, subvenient world of the very small. And as a further, critical instance, it is claimed:

All the psychological phenomena of the world supervene on all the non-psychological phenomena that are subvenient to them.

A more formal statement of the philosophical position of physicalism is this:

Given the real world R and all possible worlds,

(i) *In any world W where there is a difference from R in a mental feature, there will also be a difference in W from R in some physical feature;*
(ii) *There is a world W that is physically different from R but which has identical mental features to R.*

It is important to understand the importance of the second paragraph of this definition. Supervenience is an asymmetric relation. It affords a primacy to the fundamental, the most basic, features of reality.

But what is meant by the term 'physical'? Here the philosophical physicalist is obliged to openly avow her humility, because as a philosopher she has no means at her disposal to answer the question of what exists. She is obliged to demur to the natural scientist, the physicist, to provide the answer. The physical – what exists – is what science tells us exists.

Science may acknowledge two kinds of existent – existents supervenient on more fundamental elements, or fundamental constituents of reality beneath which there is nothing. Nobody in modern science has yet proclaimed the discovery of the fundamental level, but one day it may be reached, and if it is reached, the science will need to provide compelling evidence as to why it is fundamental and not, at least potentially, supervenient on deeper structures. It is worthwhile to consider the possibility too that it may not be possible for human science to apprehend the fundamental level – that there may be genuine limits to our knowledge – and it is interesting to consider the possibility that *there may not be a fundamental level at all.*

What has become clear is that as physics explores reality at ever smaller orders of magnitude, our ordinary concept of object, a body existing in space, becomes hopelessly inadequate. It was seen in the last chapter how the world of things existing in time and space has to be relinquished for a far more complex and subtle conceptual landscape that can seemingly only be described in an abstract mathematical language.

The physicalist claim boils down to a *prediction* in two parts – first

(i) *that physics will neither discover a psychological entity at the deeper layers that it explores;*

(ii) that physics will not posit the existence of an entity purely for the purpose of accounting for psychological phenomena.

Identifying the physicalist claim in this way, it makes clear that physicalism is a theory that could be proved false – the hallmark of a valid theory in science, and a hallmark of the kind of metaphysical theory that physicalism seeks to be.

A word about the second part of the prediction is in order. Its inclusion in the definition of physicalism is prompted by an idea in the biological sciences. The idea can be traced back to the ancient world, but in the eighteenth and nineteenth centuries there was a central ongoing dispute among biologists concerning the question of whether or not living things were fundamentally different in some crucial way from inanimate things. People who believed there to be a fundamental distinction to be drawn came to be known as *vitalists*, because the special property attributed to living things was a property of a mysterious, possibly non-material kind that gave to beings with the property the property of being alive – the vitality of life. All kinds of words and expressions came to be used to identify this special thing, the most popular coming to be a term coined as late as the early twentieth century by Henri Bergson. The *élan vital* was the name given to that property of living things that accounted for... their life. It is the positing of such a property, or entity, for the purpose of accounting for psychological phenomena that the second part of the prediction claims will not happen. Eventually vitalism was taken by the vast majority of biologists to have been refuted by empirical evidence, and was to be banished from science. But it attracted some first-order scientists, including Louis Pasteur, and of course the bridges that were finally laid linking chemistry directly with biology were far more sophisticated and complex than the attempts at a mechanistic account of life that vitalists criticised, reasonably enough, for being hopelessly inadequate for the provision of an explanation of the phenomena of life.

For reasons that will become clear, it is important to distinguish two approaches to the definition of supervenience. The concern here is with a theory that supports a physicalist outlook, and for that purpose a *global supervenience* is sufficient – *all* the psychological features of reality supervene on the underlying, subvenient base. *Local supervenience*, in contrast, seeks to link particular features of the higher layer with particular elements of the deeper layer. Such an approach makes obvious sense if we consider, again, the table. The table of our house supervenes on its constituent atoms.

But things may not be so simple or obvious in general. Take the mind of an individual person – what is the subvenient base of this mind? There are a number of answers on offer; some would say the individual's brain is the subvenient base; others might say it is the entire nervous system of the individual, and yet others would point to the entire body of the individual as the subvenient base. Even more controversially, some theorists argue that the individual mind supervenes on a broader subvenient base that extends beyond the boundaries of the individual's physical being.

There are many problems with local supervenience. The primary one is that there is no clear set of *philosophical* criteria to help in the task of choosing between these options. There is interesting and important empirical research that is exploring the link, for example, between a person's wish and the associated processes in the person's musculature and nervous system. This research is promising extraordinary technological advances, such as the ability to have the prosthetic mechanical arm of an amputee respond according to the person's wish. It is in science and technology that local supervenience can be seen to be a useful perspective. With issues of the kind science and technology struggles with, the philosophy of local supervenience is basically guesswork.

There is a deeper problem. In its philosophically most rigorous form, supervenience does not point to the supervenience of something on the layer immediately beneath it; the objects and properties of one layer *supervene on all the layers that underlie it*. This formulation is needed for a specific reason.

One feature of the layers of reality is that deeper layers bring greater complexity. For example, there is the single table, beneath which there is a structure of trillions of atoms. That structure of atoms supervenes on an even more complex array of subatomic entities. Just as the observable state of the table could arise from any one of a large number of potential states at the atomic level, so any given state at the atomic level might supervene on any one of a vast number of sub-atomic states. Now here is the vital theoretical possibility.

Suppose:

A supervenes on B, and B supervenes on C;
A is in state $a1$ and B is in state $b1$;
When B is in state $b1$, C is in state $c1$;
It is possible that C changes to state $c2$, thereby causing an alteration in the state of A from $a1$ to $a2$, while B remains unchanged in state $b1$.

An example of the possibility being suggested is this: that a change in the mental state of an individual *may not be* identified with a change at the level of the neuronal network of the person's brain, but may be related to a change at a deeper level that does not find expression in the neuronal map available to neuroscience.

Global supervenience makes a much weaker claim than local supervenience, but it is all that is required for physicalism. It is in the character of physicalism to assign to science and technology the task of exploring the specific links between the layers of reality, as when neuroscientists link the nervous system and a thought, or atomic physics accounts for the chemical bond.

Global supervenience is also weak in another sense: it does not have anything to say about the *specific nature* of the supervenient relation present between any given supervenient-subvenient dyad. Nothing is known about the comparison of the relation, on the one hand, between mind and the underlying layers, and, on the other hand, the relation between the material object and the underlying layers – except that they are both supervenient relations. However, the following section draws attention to the disturbing consequences of the strength of global supervenience.

The causal exclusion argument and the fate of causation

The situation looks paradoxical. On the one hand, an ontological theory, materialism, has been superseded by a theory, physicalism, that claims it has no tools with which to make positive ontological claims. On the other hand, physicalism claims to be the natural successor of materialism and to capture its essence. That does make it sound as though the essence of ontological materialism was not ontological, but it's not quite a bad as that. The ontological theories of the materialists can be seen as primitive attempts at science, without all the conceptual and concrete tools of modern science, and with all the innocent optimism that scientific discovery was readily available to the acute observer with a sound capacity to reason. No one could really have predicted just how obscure, paradoxical and baffling the nature of reality would prove to be.

However, the formulation of physicalism outlined above has met with a formidable argument that will expose just how radical physicalism is with respect to our everyday, cherished beliefs about our nature.

The argument is called the *causal exclusion argument*. It challenges the intelligibility of mental causation, and it can be expressed like this: if a mental event M supervenes on a physical event P, and P causes a further physical event P* on which a further mental event M* supervenes, serious doubt can be cast on the claim that M causes M*. The account at the physical level of how P causes P*, together with the supervenient relations, is sufficient to account for the occurrence of M*. The M-to-M* doesn't seem to be a genuine causal relation.

The philosopher Jerry Fodor has expressed a personal response to the problem in no uncertain terms:

> ... if it isn't literally true that my wanting is causally responsible for my reaching, and my itching is causally responsible for my scratching, and my believing is causally responsible for my saying ... if none of that is literally true, then practically everything I believe about anything is false and it's the end of the world.
>
> (Fodor, 1990, p. 156)

Here's a situation that may clarify what has worried Fodor so much. Imagine you are deciding whether or not to open the window. It *seems* as though the world has reached a fork in the road – on one possible future path the window stays shut, while on the other possible path the window is opened; furthermore, it *seems* as if you hold in your hands, in your choice, which path the future will take. However, if the subvenient base of your deliberation stands in a causal role to the subvenient base of your subsequent decision – to open or not to open – then *you* would seem to lose your central place in the story. What happens – window opened or window kept shut – can be determined without reference to the supervenient story of you and your deliberations.

Fodor's anxieties are not without foundation. It was looking very promising. In outline: the materialist relinquishes the search for what exists to the scientist, and simply predicts, on good inductive evidence, that the scientist will not posit psychological phenomena beyond those we are already familiar with. The materialist designates all phenomena, from quarks to thoughts, as things that exist but which are supervenient on some more, yet-to-be-discovered, fundamental level of reality. The materialist seems to have kept everything, including mental entities, at no cost. But in fact the cost, from Fodor's perspective, is very high. But is it so high? Is it the end of the world?

The first point to be made is that this problem was always lurking in the shadows of materialism. Epicurus, and Lucretius after him, pointed to the swerve as the key explanatory feature of the reality of atoms in the void that would account for our freedom of will. However, no such account was forthcoming. Our intuitive account of ourselves as human beings seems to be implicitly a dualist picture where the individual can stand outside the material order and make free choices. It was observed above that the identity theorists demand a radical rethink of our intuitions about mental entities. What is clear now is that *all* materialist theories demand a radical rethink of what it is to be a human being.

This chapter concludes with the outline of an account of human choices and decisions intended to save the world. To begin, in addition to the supervenience of mental entities – thoughts, desires, deliberations and so on, there must too be *actions*. To clarify this idea, imagine a human being's arm waving. This may be a simple event in the physical world, but it has become an action if it is a *waving goodbye*. An action is a physical motion of the body linked with an associated mental state.

On this basis, the deliberation about opening the window could have this analysis: you decide you want the window open; put another way, the wish for the window to be open forms in your mind. You then entertain the fantasy of the window being open and you have thoughts about the means to it being opened. Finally, on this basis, the action is undertaken to open the window.

It will not escape attention that this does not undermine the causal exclusion argument. Rather it simply provides a description of what is going on when we have the feeling that we hold the future in our hands. It does suggest that the opening of the window was based on our desire, but the roots of that desire can be traced at a sub-mental level. If the subvenient base of our minds is taken as our physical bodies, including the brain, it may be argued that this physical system 'wanted' the window open, and the mental wish expressed this. This perspective may explicate something of the identity theorist's perspective, even if identity theory must be abandoned.

Introduction

The physicalist is the materialist who has learnt the lessons of twentieth-century physics. The most basic teaching of both relativity and quantum physics is that our intuitions are of no help in the scientific quest to understand the fundamental nature of matter. The world, as conceived by modern physics, both in its vastness, as described by astrophysics and cosmology, and in its minuteness, as revealed by the quantum mechanical Standard Model of particle physics, is so strange, so immensely complex and difficult to grasp, that speculations based on ordinary ideas can contribute nothing to the painstaking practice of science.

In the face of the extraordinary progress in the human understanding of reality, as expressed in the edifice of modern physical sciences and manifested in the technological fruits of that science, the physicalist also acknowledges that the ultimate nature of reality remains something of a closed book, to scientists as well as everyone else. Chapter 6 argues that not only do we not know what the fundamental 'level' of reality is like, we do not even have good grounds for believing there is a fundamental level at all. We know that reality in itself, separate from human science and observation, is not in levels, it just is what it is.

The idea of levels arises when we realise that we can study reality at different scales, and this may or may not be a process that goes on indefinitely. As things stand at the moment, the most fundamental levels observed with the most powerful microscopes, both real and metaphorical, are described in a highly abstruse mathematics, employing concepts that have no simple description in the language of everyday life. In the physicalist perspective, reality loses none of its capacity to astonish and amaze.

Ontology

The physicalist bows before science – what exists is what science tells us exists. This, however, needs qualification. All scientific theories are subject to potential revision and abandonment, so all ontological claims are taken with a greater or lesser degree of tentativeness. It is true that we cannot seriously deny the existence of the objects of our everyday world, and few scientists would give much credence to the idea that future science will have done away with the concepts of atom, proton, neutron and electron. However, as more fundamental physics becomes more speculative our confidence in the existence of hypothesised entities such as superstrings diminishes, in the sense that a new framework could supersede the existing one with a different range of entities. However, the physicalist inherits from materialist predecessors the negative ontological claim, and the advent of physicalism makes clearer just how central to materialism the negative claim was. Only now are we able to formulate it in a more sophisticated fashion.

Materialism claims there are no gods and demons, ghosts and ghouls, spirits and fairies. It does acknowledge the existence of psychological phenomena, most obviously the sensations, thoughts, feelings and volitions of our own human experience. However, it recognises all psychological phenomena as supervenient on – wholly dependent for their existence on – a non-psychological base. That is to say, physicalism *predicts that physics will not hypothesise the existence of spiritual or psychological phenomena for the purpose of explaining psychological phenomena.* In this way physicalism is a *refutable theory.* If physics were to hypothesise an entity with demonstrably psychological properties in order to account for psychological phenomena, that hypothesis would be incompatible with physicalism, and if it proved successful and became adopted as out best physics then physicalism would be refuted.

Epistemology

The epistemological perspectives are carried forward from materialism to physicalism. The materialist has always grounded knowledge in the empirical study of reality. However, hand in hand with developments in science, the scientific method has undergone profound refinement and elucidation. In particular, higher and higher critical standards have been set for the acceptance of theories, and many theories are only ever held with some degree of tentativeness. Too many shocks and surprises in the investigation of the world have left scientists wary of claims of certainty.

The connection between the scientific method and materialism and physicalism is complex. On the one hand, much about science as it is now done, especially in biology and chemistry, presupposes physicalism, and the origins of modern science lie in the repudiation of immaterial essences and forms. On the other hand, physicalism is a falsifiable theory, and in virtue of this falsifiability physicalism claims legitimacy as a theory about reality. In principle science could disprove physicalism. However, the physicalist outlook is so deeply embedded in the practice of theoretical physics, that if the point were reached where it was felt necessary to postulate the existence of a mental or spiritual entity to provide an explanation of some phenomenon, science would have reached a stage in its development so revolutionary, demanding such a profound change in perspective, that everything would be up for question, including the fundamental epistemological outlook established in the seventeenth century.

The physicalist critique of other ontologies

There is greater humility in physicalism than materialism. Indeed, Epicurus and Lucretius are all too sure of themselves for modern sensibilities. Nonetheless there is probably a fundamental division amongst people with regard to the question of whether or not reality has a spiritual dimension. Of course, there may be a large constituency of people who remain neutral, agnostic, on the question. But for the others, there seems to be something almost ungraspable about how anyone could hold the opposite view to their own. For much of history, the idea of there not being a god has been thought by the vast majority of people whose views have been expressed in written language as *absurd*. The majority of adults living now also ascribe to a belief in god. For the materialists, taking on board the strange news from twentieth-century physics did not involve a heart-searching with regard to the non-existence of the spirit for very long. It became clear that in the extraordinary description of the worlds of the very large and the very small given by modern physics, spiritual things were not being offered or suggested as part of the story.

The physicalist worldview

However, the physicalist has discarded the social militancy that became attached to materialism in the nineteenth century. It is a philosophical theory; it has an attitude to science and ontological claims; it ascribes to a scientific methodology; it makes predictions about future theory. It is not a social theory, nor part of a political programme. It meets its opponents in philosophical debate, not in physical combat. Many physicalists will adhere to the principles of the radical enlightenment of d'Holbach and others, but this is true of vast swaths of the population now.

In more general terms, physicalism is aligned with a belief in rationality and its importance in human society. This transcends differences in philosophical theories and perspectives. As considered in the final chapter, rationality faces many challenges in the modern world, and its defence is not straightforward. There is no consensus about the foundations of rationality, and the enemies of rationality, as understood by the philosophical community, may choose physical combat over discussion. But that is not a new phenomenon. The pen is mightier than the sword – sometimes.

Eight

At the time of writing, in the second decade of the twenty-first century, the state of the world, and of human civilisation, can look quite paradoxical and quite precarious. In certain ways, the state of the contemporary world has interesting ramifications for the question of physicalism. Take, by way of evidence, two articles from *New Humanist*, the journal of the Rationalist Association, in the summer of 2016. The first concerns an unorthodox Christian minister:

> Can a Christian church employ an atheist minister? That is the question currently being examined by the United Church of Canada. Gretta Vosper, the minister who has led the congregation at West Hill in Toronto since 1997, affirms: 'I do not believe in a theistic, supernatural being called God. I don't believe in what I think 99.99 percent of the world thinks you mean when you use that word.'
>
> Vosper's sermons are light on scripture, often eschewing the word 'God' altogether and focusing instead on general moral teachings. It appears to be working. Her 100-strong congregation has not faced the declining attendance rates of other churches.

When she was ordained as a minister in 1993, Vosper was asked whether she believed in God, the Father, the Son and the Holy Spirit. She said yes, speaking metaphorically. Almost a decade later, frustrated by the archaic language and imagery of the Bible, she delivered a sermon deconstructing the idea of God. Rather than cracking down, the United Church of Canada – known as a progressive and broad-minded denomination – encouraged her to push the boundaries. For years, Vosper described her position with a series of linguistic contortions, labelling herself a non-theist and a theological non-realist.

The tipping point came in 2013, when she began to identify herself publicly as an atheist. She has said in interviews that this was not the result of any change in her belief system but a wish to show solidarity with Bangladeshis suffering persecution. Her words caused outrage in some quarters, raising questions about why exactly an atheist was preaching in a church at all – even one that prides itself on its progressive values.

Vosper's Twitter bio claims that she is 'irritating the church into the twenty-first century'. Whether she succeeds in her mission to carve out a space within religious institutions for those who do not believe in God remains to be seen. With her legal appeals to prevent an official United Church review rejected, her days at the pulpit may be numbered.

(Shackle 2016, p. 12)

If atheist priests are a hard act to follow, have a look at this excerpt about an unfortunate Russian atheist:

From materialism to physicalism

Still, it's difficult not to be taken aback at the sheer scale of change in Russia over the past few decades. In 1961, after becoming the first person to leave the Earth's atmosphere, the Soviet cosmonaut Yuri Gagarin was reported by state media to have said: 'I travelled into space but I didn't see God there.' (Whether or not he actually uttered these words is a matter of debate.)

In 2016, similar atheist statements could land you behind bars. In Stavropol in southern Russia, a 38-year-old blogger named Viktor Krasnov was charged earlier this year with 'insulting the feelings of religious believers'. His crime? Writing 'There is no God' during a heated online discussion. He faces up to a year in prison. Even before his case came to trial, Krasnov was forced to undergo a monthlong psychiatric evaluation by a judge, who told him: 'No one in their right mind would write anything against Orthodox Christianity and the Russian Orthodox Church.' Lenin must surely be spinning in his glass coffin.

(Bennetts 2016, p. 37)

Spinning indeed. And the physicalists are surely scratching their baffled heads. Promoting concern at a more visceral level is what has come to be known as the age of 'post-truth'. The problem can be seen as an undermining of the epistemological precepts of materialism and physicalism. The last chapter ended with thoughts about the confrontation of rationality and irrationality, and the ways this conflict may be enacted, and noted that the pen is, sometimes, mightier than the sword. It should be noted also that sometimes the pen is poisoned.

The conservative Evangelical Protestant tradition in America seems to have extensive influence in American politics in the second decade of the twenty-first century. Molly Worthen has written in the *New York Times* of strands of thinking within this tradition (NYT, 13/4/17). At the centre is the Bible, towards which there are two attitudes. One, called presuppositionalism, rejects science's right to challenge the Bible. Another, not incompatible with that viewpoint, sees the Bible as *inerrant*, as both 'supernatural and scientifically sound', and the basis of a 'biblical', or 'Christian', worldview. This worldview is at odds with the theory of evolution and the science of climate change. For present purposes, the important feature of this tradition is the fact that significant numbers of people are willing to accept the assertion that a text has a transcendentally special authority without any critical scrutiny of the grounds on which this claim is made.

There is so much paradox to acknowledge, and sadly so many places where reasoned debate is not an option. The march of technological progress, based on a scientific conception of the world that itself is founded on a materialist-physicalist epistemology, has accelerated and shows no sign of slowing down. In the developed world, many religious bodies face declining numbers of followers and are struggling to hold on to their influence and authority, while in vast tracts of the developing world various religious world views, often accompanied by a savage morality, have millions of people in their thrall. The old fear and loathing the religious authorities have for materialist and atheistic perspectives is expressed in the laws of a not-inconsiderable number of countries where to uphold, sometimes even to believe, such things is punishable by death.

Conflict between the religions can also exhibit extreme violence. An atheist once remarked that religious wars are people killing each other over who has the best imaginary friend. But whenever avowed atheists have achieved state power they have exhibited a capacity for violence and repression no less savage than the worst of the religious authorities in the past 3000 years. At the political level, the present century has inherited from the past one a profound distrust of materialist authority. At the time of writing the major avowedly atheistic regimes can boast few international admirers. Notwithstanding the decline in religiosity in the developed world, it would be very unlikely that an avowed atheist would be elected President of the United Sates, or appointed Prime Minister of the United Kingdom.

So what place, and what role, for physicalist-materialism in this turmoil? First, its primary place is wherever philosophy is conducted. In modern times that is likely to be in university departments of philosophy, but there is always the hope that philosophy is not exclusively confined to universities. Here there are, evidently, many potential lines of research. To name the core ontological concerns: first, the concept of supervenience may undergo further analysis and refinement. As described earlier, it is, seen from different perspectives, both extremely weak and extremely strong as a doctrine. Further reflection on the circumstances in which the qualitatively new phenomena emerge may provide further insight into this aspect of the non-smooth evolution of the universe. Second, the consequences of a materialist physicalism based on supervenience for the nature of *causation* demands much further attention. A radical rethinking of the concept of causation is required in the natural sciences, and something little short of a revolution is threatened in relation to the assessment of our agency as human beings. This links to the third line of research, perhaps the most pressing of all for our self-conception: the nature of free will in a world that sees the mental realm as supervenient on the physical realm. Of course, the dispute with non-materialist-physicalist theories will go on. In summary, the physicalist has a place in the consideration of central issues of epistemology and metaphysics in general, and philosophy of mind and philosophy of science in particular.

At a more speculative level, the doctrine of supervenience may give licence to a legitimate sense of wonder about the natural world. Materialism has been accused of banishing all the wonder associated with a religious view of the world, but this seems evidently wrong. In questioning the evidence for the mystical, nothing is denied about the strangeness of reality as described by physics, chemistry and biology.

To speculate even further, the moral world itself may come to be seen as a supervenient phenomenon in the diachronic development of the universe. That morality is in some sense a supervenient phenomenon has already been argued in the tradition of analytic ethics. Hare observes that two people cannot be alike in every way except that one of them is a good person and the other is not. If the supervenience of the good were somehow latent in the physical universe, physicalism may find itself exhibiting unexpected aspects.

Whatever avenues physicalism follows within philosophy, within science the battle is already won – and, barring the collapse of civilisation, safely won. There would seem to be overwhelming evidence that the only research into the nature of the world that bears demonstrable fruit is based

on the guiding principles of science – observing, hypothe-
sising, experimenting, testing. Notwithstanding this, many
people live by a split in their mind between their materialist
selves and their non-materialist selves. There is often great
discomfort, even pain, in accepting a materialist conception
of the human being, and this split in the mind attempts to
circumvent that pain. The attitude recalls the way many of
the scientists of the early modern period managed the task
of pursuing science while appeasing their inner or outer
priest.

Within the sociopolitical sphere, while we have to
acknowledge that no ethical or political stance is implied by
materialism, it has throughout its history, until the twentieth
century, been associated with humanism, in one form or
another. At its zenith in the Radical Enlightenment,
expressed most clearly in d'Holbach's *System of Nature*, it can
be seen as the ontological underpinning of the ethics that has
come to enjoy a near-consensus amongst contemporary non-
religious traditions, as well as many religious traditions as
well – tolerance; non-discrimination on grounds of race,
sex, gender or sexual orientation; separation of Church and
State; and freedom of thought and expression. From this
point of view, it is clear how the materialist stands in opposition to religious fundamentalists in the struggles of the contemporary world. Whatever the historical links with the
Enlightenment and humanism that materialism has, the
materialist physicalist will have to find the grounds for ethics
beyond the core philosophical beliefs discussed in this book.
However, the materialist has at least two crucial roles: first to
challenge fundamentalist religion and its pretensions to
knowledge and authority, and second to show that materialism as a worldview has no necessary connection with the

repression and violence associated with states that espouse a materialist-atheistic ideology.

From this point of view, the materialist has only a limited offering to the ancient philosophical concern of how to live. She has a rational argument for abandoning fear of death and the afterlife, the wrath of gods or the priests' threats of retribution to come. However, she has no resources, qua materialist, to challenge fear, or any other emotion, nor does she have any guidance concerning the superiority of one way of life over another. Though to dispel the supernatural may be a good place to start.

A plea for tolerance, sensitivity, humility and respect

This last comment requires a caveat. When materialism takes part in a public discussion beyond the philosophy department, it is important to bear in mind how much emotional investment many people have in non-materialist worldviews. Beyond individual investment, many communities depend to a significant degree on non-materialistic, typically religious worldviews, for their social cohesion and functioning. The ruthless hostility to materialist ideas in the sixteenth and seventeenth centuries was largely to do with its threat to established authorities, but it also seemed to some reasonable people that it threatened the very fabric of society. Recall again d'Holbach's doubts about exposing materialist thought to the mass of the population.

Whatever your views as a materialist on those religious worldviews, and on the manner of social functioning that the views support, an intolerant, insensitive or disrespectful attitude in the debate is liable to result in destructive consequences.

There is no advancement for the materialist worldview either in being instrumental in a successful assault on a society's foundational perspectives, or in provoking violent opposition. With this in mind, it is salutary to remember two things. Modern physicalist materialism prides itself on its refutability. It may be false, and it may be proved to be demonstrably false by future advances in human knowledge. However convinced physicalists are or feel, the falsifiability of the worldview demands a degree of humility. Second, much of the epistemological foundations of the materialist perspective focuses on conditions of adequacy for evidence cited to support an argument or theory. Many people have had experiences that do not meet these criteria but which, for them individually, provide convincing, sometimes incontrovertible evidence of the existence of that which materialism denies. The doubts the materialist has about the implications of these experiences do not justify disrespect for the individuals that have them. At this stage in human history, it is important to remember that, taking the human species as a whole – both those alive now, and all those who have ever lived – the materialists are very definitely in a minority. About such matters there can be no absolute certainty.

Bibliography

Items marked with an asterisk are not referred to in the text.

*Baggini, J. (2003) *Atheism: A Very Short Introduction*. Oxford: Oxford University Press.

Bennetts, M. (2016) 'The Resurrection of Belief'. *New Humanist*, Summer: 34–7.

Bentley, R. (1838) *Sermons Preached at Boyle's Lecture, etc.*, ed. A. Dyce. London: Francis Macpherson.

Bhattacharya, R. (2011) *Studies on the Carvaka/Lokayata*. London and New York: Anthem Press.

Brown, R. & Ladyman, J. (2009) 'Physicalism, Supervenience and the Fundamental Level'. *Philosophical Quarterly* 59, no. 234: 20–38.

Bullivant, S. & Ruse, M., eds. (2013) *The Oxford Handbook of Atheism*. Oxford: Oxford University Press.

Carr, B. & Mahalingam, I., eds. (1997) *Companion Encyclopedia of Asian Philosophy*. London and New York: Routledge.

Crane, T. (2017) 'How We Can Be'. *Times Literary Supplement* 5956, 7–8.

Diogenes Laertius. (2015) *Complete Works*. Delphi Classics. www.delphiclassics.com.

Fodor, J. (1990) *A Theory of Content and Other Essays*. Cambridge, MA: MIT Press.

Foglia, M. (2014) 'Michel De Montaigne'. In *The Stanford Encyclopedia of Philosophy*, ed. E.N. Zalta (Spring 2014 Edition). https://plato.stanford.edu/archives/spr2014/entries/montaigne/.

Frazier, J. (2013) 'Hinduism'. In *The Oxford Handbook of Atheism*, eds. S. Bullivant & M. Ruse. Oxford: Oxford University Press, pp. 367–82.

Gain, D. (1969) 'The Life and Death of Lucretius'. *Latomus* T.28, no. Fasc. 3, pp. 545–53.

Gokhale, P. (2015) *Lokayata/Carvaka: A Philosophical Inquiry*. Oxford: Oxford University Press India.

Gooch, T. (2016) 'Ludwig Andreas Feuerbach'. In *The Stanford Encyclopedia of Philosophy*, ed. E.N. Zalta (Winter 2016 Edition). https://plato.stanford.edu/archives/win2016/entries/ludwig-feuerbach/.

Gottlieb, A. (2016) *The Dream of Enlightenment*. London: Allen Lane.

Greenblatt, S. (2012) *The Swerve*. London: Vintage Books.

Herrick, J. (1985) *Against the Faith*. London: Glover & Blair Ltd.

Hume, D. (2007) *An Enquiry Concerning Human Understanding*, ed. P. Millican. Oxford: Oxford University Press.

Israel, J. (2002) *The Radical Enlightenment: Philosophy and the Making of Modernity 1650–1750*. Oxford: Oxford University Press.

Jammer, M. (1999) *Einstein and Religion*. Princeton, NJ: Princeton University Press.

Joshi, L. (1966) 'A New Interpretation of Indian Atheism'. *Philosophy East and West* 16, no. 3–4: 189–206.

Kenny, A. (2004) *Ancient Philosophy: A New History of Western Philosophy*, volume 1. Oxford: Clarendon Press.

Kirk, G. & Raven, J. (1964) *The Presocratic Philosophers*. Cambridge: Cambridge University Press.

Lucretius. (1997) *On the Nature of the Universe* (De Rerum Natura), translated by R. Melville. Oxford: Oxford University Press.

Madhava. (1978) *The Sarva Darsana Samgraha of Madhava*, ed. V.S. Abhyankar, translated by E.B.Cowell & A.E. Gough. Pune: Bhandarkar Oriental Research Institute.

Marx, K. (1844) *Economic and Philosophic Manuscripts*. www.marxists.org/archive/marx/works/download/pdf/Economic-Philosophic-Manuscripts-1844.pdf.

Marx, K. (2000) *Selected Writings*, ed. D. McClellan. Oxford: Oxford University Press.

McEvilley, T. (2002) *The Shape of Ancient Thought*. New York: Allworth Press.

Nadler, S. (2013) *A Book Forged in Hell*. Princeton, NJ: Princeton University Press.

Newton, I. (1710–) 'A Short Schem of the True Religion'. Keynes Ms. 7, King's College, Cambridge. www.newtonproject.ox.ac.uk/catalogue/record/THEM00007.

Newton, I. (2016) *The Principia: The Authoritative Translation: Mathematical Principles of Natural Philosophy*, translated by I. Cohen, A. Whitman & J. Budenz. Oakland, CA: University of California Press.

O'Connor, J., ed. (1969) *Modern Materialism: Readings on Mind-Body Identity*. New York: Harcourt, Brace & World, Inc.

Rahe, P. (2007) 'In the Shadow of Lucretius: The Epicurean Foundations of Machiavelli's Political Thought'. *History of Political Thought* 28, no. 1: 30–55.

Bibliography

Riepe, D. (1964) *The Naturalistic Tradition in Indian Thought*. Delhi: Motilal Banarsidass.

Rovelli, C. (2014) *Seven Brief Lessons on Physics*. London: Penguin Books.

*Russell, B. (1947) 'Am I an Atheist or an Agnostic?' http://scepsis.net/eng/articles/id_6.php, copyright 2005Scepsis.net.

Shackle, S. (2016) 'Atheist Minister'. In Witness section, *New Humanist*, Summer: 12.

Shortt, R. (2016) 'At the Prow of History'. *Times Literary Supplement* 5933, 3–5.

Spinoza, B. (1967) *Ethics*, ed. J. Gutmann. New York and London: Hafner Publishing Company.

Spinoza, B. (2007) *Theological-Political Treatise*. Cambridge: Cambridge University Press.

Tiehen, J. (2018) Physicalism. *Analysis* 78, no. 3: 537–51.

Werner, K. (1997) 'Non-Orthodox Indian Philosophies'. In *Companion Encyclopedia of Asian Philosophy*, eds. B. Carr & I. Mahalingam. London and New York: Routledge, pp. 103–18.

*Whitmarsh, T. (2017) *Battling the Gods: Atheism in the Ancient World*. London: Faber and Faber.

*Wolfe, C. (2016) *Materialism: A Historico-Philosophical Introduction*. Cham: Springer. (Authors' note: despite the title, this book is not written for the general reader. It is a work of academic scholarship for a readership of professional philosophers and advanced students.)

Worthen, M. (2017) 'The Evangelical Roots of Post-Truth'. *New York Times*, *International Edition*, 13 April 2017.

Name Index